MW01490242

Fostering Communities of Transformation in STEM Higher Education

A Multi-institutional Collection of DEI Initiatives

Fostering Communities of Transformation in STEM Higher Education

A Multi-institutional Collection of DEI Initiatives

Jonathan S. Briganti
Jill C. Sible
Anne M. Brown

Access the free downloadable PDF:
https://doi.org/10.21061/fosteringcommunities

Virginia Tech Publishing
Blacksburg, VA

Copyright © 2024 by Jonathan S. Briganti, Jill C. Sible, Anne M. Brown

First published in 2024 by Virginia Tech Publishing

Virginia Tech Publishing
University Libraries at Virginia Tech
560 Drillfield Drive
Blacksburg, VA 24061
USA

This work is licensed under the Creative Commons Attribution 4.0 International License. To view a copy of this license, visit https://creativecommons.org/licenses/by/4.0/deed.en or send a letter to Creative Commons, PO Box 1866, Mountain View, CA, 94042, USA.

Note to users: This work may contain other components (e.g., photographs, figures, or quotations) not covered by the license. Every effort has been made to identify these components but ultimately it is your responsibility to independently evaluate the copyright status of any work or component part of a work you use, in light of your intended use. Brand names, trademarks, and service marks in this book are legally protected and should not be used without prior authorization by their owners. Their inclusion in this book is for analytical, illustrative and pedagogical purposes only.

Cataloging-in-Publication Data
Name: Briganti, Jonathan S. | Sible, Jill C. | Brown, Anne M. – authors.
Title: Fostering Communities of Transformation in STEM Higher Education: A Multi-institutional Collection of DEI Initiatives | Jonathan S. Briganti, Jill C. Sible, and Anne M. Brown.
Description: Blacksburg, VA : Virginia Tech Publishing, 2024. | Includes bibliographical references. | Summary: "This scholarly work examines transformative initiatives from Virginia Tech, Radford University, Trinity Washington University, and Towson University, showcasing their role as catalysts in cultivating inclusive excellence across diverse STEM disciplines." – Back cover.
Identifiers: ISBN: 978-1-957213-84-2 (PDF) | ISBN: 978-1-957213-83-5 (epub) | ISBN: 978-1-957213-82-8 (print) | DOI: https://doi.org/10.21061/fosteringcommunities
Subjects: Science--Study and teaching (Higher)--United States--Textbooks. Technology--Study and teaching (Higher)--United States--Textbooks. Engineering--Study and teaching (Higher)--United States--Textbooks. Mathematics--Study and teaching (Higher)--United States--Textbooks. Multicultural education--United States--Textbooks.

Cover design: Catherine Freed

CONTENTS

FOREWORD: CURIOSITY, COURAGE, AND COMMUNITY

DAVID J. ASAI

All persons deserve the opportunity to experience the process and language of scientific thinking. A key element of social justice, STEM literacy is the gateway to personal agency and empowerment in every aspect of our lives, including healthcare, the environment, energy, artificial intelligence, communications, and transportation. Just as important is that all of us, regardless of our profession, be able to apply the critical thinking skills that are at the core of the scientific process as we make decisions about ourselves, our families, and how we participate as citizens of our planet.

The undergraduate years present the best opportunity to engage large numbers of students in learning the process of science. Every year in the U.S., about three million students enter college for the first time, and most of them will take at least one course in a STEM discipline as part of a general education requirement. And nearly one million first-time students plan to major in a STEM discipline.

More than half of the students who enter planning to study STEM do not leave with a STEM degree. Much more troubling are the gaping economic, generational, and racialized disparities in the success of students in STEM. Students who are first in their family to attend college, who begin at a community college, or who identify as belonging to populations excluded because of ethnicity or race are far less likely to complete a STEM degree than their counterparts who come from families with a parent who has a bachelor's degree, who begin at a four-year school, or who are white or Asian.

For decades, our community's response has mainly been to "fix the student." We have created an impressive array of student-centered interventions including outreach to K-12 students, pre-college summer bridge programs, special tutoring and advising, undergraduate research opportunities, post-baccalaureate programs, and "minority" supplements to federal research grants. These interventions can help the assimilation of the students who participate in the programs, but they almost never change the educational system. The continuation of any positive outcomes depends on the continuous influx of money, energy, and time. This is a fragile and unsustainable strategy, because when the financial support ends, so too does the intervention and the system reverts back to where it was before. Student-centered programming is not a viable strategy for lasting culture change.

EXCELLENCE

"Excellence" is a word frequently used but infrequently understood: we all want to be it but we can't quite explain what it is. Look up online the strategic plan of your favorite college or university and then search the text for "excel." Almost certainly you will find the word or its derivatives "excellent" and "excellence." But does the document explain what they mean by "excellence?"

We often think excellence is an objective measure of quality – for example, "Maria is an excellent student," or "University X is the home of unprecedented excellence in research and teaching." But excellence is neither immutable nor absolute. According to the Merriam-Webster online dictionary "excellent" is defined as first-class and superior, and to "excel" is to be superior, to surpass others. Thus, "excellence" is not a stand-alone measure of quality; instead, it is a claim that one is better than everyone else. Unfortunately, we can't all be Number One, despite what we might claim in our strategic plans.

Rather than vying for empty superiority, an educational institution should aspire to creating a learning environment that encourages and nurtures what Eric Weiner refers to as "genius," born from a convergence of disorder, discernment, and diversity (Weiner, 2016). STEM is all about problem-solving, and diversity is essential to successful problem-solving. As Scott Page demonstrates, innovative and creative problem-solving emerge when we embrace divergent thinking; a group's ability to solve difficult problems increases as its diversity increases (Page, 2007). But diversity without inclusion is an empty gesture. Diversity will have its positive impact only when the system ensures that every person feels that they belong and will be successful.

INCLUSIVE EXCELLENCE

That student success depends on where a student comes from rather than where they want to go is the greatest challenge in American higher education. Instead of only pursuing interventions aimed at fixing the students, it is time for a new approach, and this movement is called Inclusive Excellence.

The term "inclusive excellence" was introduced in 2002 by Alma Clayton-Pedersen and colleagues at the Association of American Colleges and Universities (AAC&U) (Clayton-Pedersen & Clayton-Pedersen, 2008). Inclusive excellence shines a light on the truth that inclusion and excellence are inextricably intertwined: there is no "excellence" without inclusion. Inclusive excellence is a dynamic stance and not an endpoint: individuals and institutions committed to inclusive excellence continuously and forever strive to create structures and behaviors centered on equity (The Inclusive Excellence Commission, 2018).

The HHMI Inclusive Excellence initiative challenges colleges and universities to think differently about student success by pivoting from student-centered activities to a commitment to increase their institutional capacity for inclusion. The Inclusive Excellence grants support faculty and staff learning and professional development which, in turn, will lead to new ways to teach science, changes in the curriculum and course content including laboratory courses, better ways to evaluate and reward teaching effectiveness and inclusivity, and greater transparency and attention to inclusion when a student transfers from one institution to another.

Inclusive Excellence is based on three foundational values: curiosity, courage, and community.

Let us be **curious** about institutional culture and how it affects our students, identifying the institutional barriers to inclusion, and learning the skills of inclusion which include listening to understand and guided reflection.

Let us be **courageous**, calling out the problems of our current culture, confronting head-on the barriers to inclusion, and engaging in the hard work of dismantling – and not simply by-passing – those barriers to inclusion. And let us have the courage to try, to sometimes fall short, and to rise up and try again.

Let us embrace the power of **community** in which no one person or one institution has all the answers. By sharing with humility what we've tried and what happened – successful or otherwise – we all benefit because the ideas will be remembered and the good ideas amplified.

This book presents the stories of four universities that were organized into an Inclusive Excellence Peer Implementation Cluster (PIC). Before the PIC was formed, the four likely did not know one another well, nor had they collaborated much. The purpose of the PIC is simple: to provide a safe forum in which the members can share their ideas, talk about what they're trying on their respective campuses, and serve as critical friends to one another.

The fifteen chapters comprising this volume are the stories of campus champions, each of whom is pursuing in their way the stance of Inclusive Excellence. Each exploration of an idea represents a contribution to overall culture change. The ideas summarized here include course re-design, organization of the curriculum and sequencing of courses, creating ways for students to engage in research, instructor professional development in the skills of inclusion, effective mentoring, and leveraging institutional change. As we read their stories, let us see how each is driven by curiosity, recognize the courage required to try new ideas, and appreciate that these progress reports are being shared with the community. Let these stories inspire us to join the movement of Inclusive Excellence.

– David J. Asai, the Senior Director for Science Education at the Howard Hughes Medical Institute.

REFERENCES

Clayton-Pedersen, A., & Clayton-Pedersen, S. (2008). Making Excellence Inclusive, in Education and Beyond. *Pepperdine Law Review 35(3)*. http://digitalcommons.pepperdine.edu/plr/vol35/iss3/3

The Inclusive Excellence Commission (2018). *Excellence: A Renewed call for Change in Undergraduate Science Education, Association of American Colleges and Universities*. https://www.aacu.org/publication/excellence-a-renewed-call-for-change-in-undergraduate-science-education

Page, S. E. (2007). *The Difference*. Princeton University Press.

Weiner, E. (2016). *The Geography of Genius*. Simon & Schuster.

ACKNOWLEDGMENTS

EDITORIAL

Graphics and Cover Design: Catherine Freed

Citation Editorial Support: Amanda MacDonald

Layout, Accessibility, and Typesetting: Truitt Elliott

All contributors provided peer review during the review process.

This program is supported in part by a grant to Virginia Tech from the Howard Hughes Medical Institute through the Inclusive Excellence Grant.

ABOUT THE EDITORS AND CONTRIBUTORS

In the ever-evolving landscape of STEM education, the call for inclusivity and equity has never been more resounding. In the following chapters of "Fostering Communities of Transformation in STEM Higher Education: A Multi-institutional Collection of DEI Initiatives" we are guided by the belief that participation in the scientific process is not only a fundamental right but also a cornerstone of social justice that paves the way for personal agency and empowerment across diverse spheres, from healthcare to artificial intelligence, from energy to environmental sustainability, and throughout the breadth of STEM.

Much like tending to a garden, in STEM education we recognize that we must both serve the individual and the whole, understanding that all individual elements are essential to the flourishing ecosystem of knowledge. Our book highlights transformative initiatives from Virginia Tech, Radford University, Trinity Washington University, and Towson University that serve as the fertilizer nurturing the garden of inclusive excellence in STEM.

Within the pages of our book, you will discover fifteen chapters that delve into the roots of inclusive excellence. These chapters are the stories of campus champions, each pursuing the stance of Inclusive Excellence in their own way. These explorations encompass an array of initiatives, from course redesign to curriculum revision, from increasing students' access to research to building instructors professional skill in inclusion, from effective mentoring to institutional change.

In this edited collection, we place a spotlight on the critical role of students as participants in reshaping the STEM landscape. They are not passive recipients but dynamic agents of progress, akin to the role any one plant plays in promoting a strong supportive ecosystem necessary for a thriving and diverse garden.

Through these initiatives that impact students, faculty, and all members of the university ecosystem, we seek to inspire you to join the movement of Inclusive Excellence, where curiosity, courage, and community converge to create a brighter future in STEM education. Adapt our ideas, materials, and lessons learned to aid your own STEM programs. "Fostering Communities of Transformation in STEM Higher Education: A Multi-institutional Collection of DEI Initiatives" is not just a book; it's a call to action, an exploration of possibilities, and a testament to the power of unity in fostering a more inclusive and equitable academic community.

THE EDITORS

Jonathan S. Briganti works in the Virginia Tech University Libraries as the manager of the DataBridge program, which trains undergraduate students from across all disciplines in applied data science and consults with partners across and beyond campus to improve the quality of their data and its outputs. He received both his BS and MS from Virginia Tech and has since found passion in creating open-source educational resources and accessible research environments to bring a more engaged and diverse pool of researchers to the table.

Jill C. Sible serves as Associate Vice Provost for Undergraduate Education and Professor of Biological Sciences at Virginia Tech where she has worked since 1998. She is the Program Director of Virginia Tech's Inclusive Excellence project, which has empowered science faculty and departments to apply a learning mindset and data-informed approach to changing classes, curricula, and culture to be inclusive of all students, especially those historically marginalized in science and higher education.

Anne M. Brown is an Associate Professor and Associate Director in Research and Informatics under University Libraries, Virginia Tech and is an Affiliate Professor in the Biochemistry Department. As a computational biochemist, her research focuses on computer-aided drug discovery and the aggregation process of amyloids. She is committed to undergraduate research and outreach. Today, she continues in her lifelong mission to create and expand opportunities for students of all backgrounds and provide them with mentorship to facilitate their success.

CHAPTER CONTRIBUTORS

Patricia G. Amateis graduated with a B.S. in Chemistry Education from Concord University in West Virginia and a Ph.D. in Analytical Chemistry from Virginia Tech. She was on the faculty of the Chemistry Department at Virginia Tech for 38 years, until her retirement in September 2022. She taught General Chemistry and Analytical Chemistry and served as the Director of General Chemistry and as the Director of Undergraduate Programs. She taught thousands of students during her career and was awarded the University Sporn Award for Introductory Teaching, the Alumni Teaching Award, the Jimmy W. Viers Teaching Award, and the William E. Wine Award for a history of university teaching excellence.

John Bernard Gonzalez Jr. is an applied mathematician at the United States Department of Defense. He earned his Ph.D. and M.S. in mathematics from Northeastern University, and S.B. in mathematics from the Massachusetts Institute of Technology. He served as guest co-editor with his wife Diana, of a 2022 special issue of The Mathematics Enthusiast focused on course-based undergraduate research experiences. He and Diana have co-authored manuscripts on data science projects related to figure skating scoring and presented this work at mathematics conferences organized by the Association of Mathematical Society and the Mathematical Association of America. Outside of work, he has earned USA Dance national adult ballroom titles at the championship level with his wife Diana in 2021, 2022, and 2023. Their biggest lifelong social justice passion project is raising an internationally adopted child.

Sharon Blackwell Jones is an Assistant Teaching Professor of Counselor Education at Wake Forest University. She a Licensed Professional Counselor and CEO of Jones Consulting and Counseling where she provides DEI Trainings, Diversity Consulting as well as Counseling Services. For the past 20 years she has worked in Institutes of Higher Education across the South East United States and K-12 Schools both teaching and providing training for both faculty and Staff in the areas of Appreciating Cultural Differences and building common ground. Dr. Blackwell Jones graduated from Howard University with a Bachelors Degree in Psychology. She went on to earn a Masters Degree in Counseling from The George Washington University and holds a Ph.D. from Penn State University in Counseling Psychology. She lives in Greensboro and St. Simons Island.

Diana S. Cheng is a Professor in the Department of Mathematics at Towson University, Maryland, and serves as the Graduate Program Director for the Masters of Science degree program in mathematics education. She earned her Ed.D. in Curriculum and Instruction from Boston University, M.Ed. in Learning and Teaching from Harvard Graduate School of Education, and S.B. in mathematics from the Massachusetts Institute of Technology. In the past year, her work on justice-centered pedagogy was supported by the Mathematical Association of America for community outreach activities, Howard Hughes Medical Institute for undergraduate courses for pre-service teachers, and the National Science Foundation for development of graduate courses and professional learning experiences for in-service teachers. She hopes to help teachers shape the world to be a more inclusive environment for all and to empower students with a sense of agency to enact change.

Kimberly Corum is an Associate Professor in the Mathematics Department at Towson University. She earned her B.A. in English Language and Literature from the University of Virginia. As a Math for America fellow, she earned her M.A.T. in Mathematics from Bard College and spent four years teaching mathematics and science at a small public high school in the South Bronx, NYC. She then returned to the University of Virginia, where she earned her Ph.D. in Mathematics Education. Her research interests include mathematics teacher preparation, mathematical problem solving, and the use of technology in the teaching and learning of mathematics.

Cynthia A. DeBoy is a Professor of Biology at Trinity Washington University. She has been the project director for the HHMI Inclusive Excellence grant with which Trinity implemented a curriculum revision to include a course-based mentor structure and faculty learning community initiatives, all aimed to support student belonging and success. Cynthia has taught a variety of courses at Trinity including newly designed molecular biology and microbiology courses. With previous research focused in neuroimmunology, Cynthia has taken a keen interest to expand research related to creating inclusive learning environments that support student belonging, self-efficacy, learning and success in science.

Erin Drolet is a PhD student in the Biochemistry Department at Virginia Tech. She is currently a Graduate Teaching Scholar through the College of Agriculture and Life Sciences, and is researching how to improve the universal accessibility of STEM laboratory courses.

Erica Echols-Miller serves as the Evaluation Manager for the National Institute for Synergy Evaluation Institute at the University of Tennessee, Knoxville. Her research interests include evaluation in higher education with special interest in diversity, equity, and inclusion impacts on student success and building diverse workforces. Erica graduated with her Ph.D. in Educational Psychology and Counseling with a concentration in Evaluation, Statistics, and Methodology from the University of Tennessee, Knoxville. She also earned her B.S. in Chemistry from North Carolina A&T State University and an M.S. in Environmental Science and Policy from the University of South Florida St. Petersburg.

Deborah J. Good is a first-generation college student raised in upstate NY in a small town of under 10,000 residents and by lower income parents. My high school guidance counselor advised me to go into medical technology since I liked science, but after doing undergraduate research both at SUNY-Fredonia and during a summer research internship at the Lovelace Inhalation Toxicology Research Institute in Albuquerque, NM, I knew I wanted to become a professor and help others discover research as well, as I understand first-hand how undergraduate research and involvement in experiential learning can change one's trajectory. I credit my professors at SUNY College at Fredonia for allowing me to do undergraduate research in their laboratories, and Sigma Xi, the Scientific Research Honor Society for recognizing me as a young scientist and inducting me into their organization. I have run four undergraduate research programs -USDA Scholars, HNFE Scholars, TOUR Scholars, and IE Scholars as a faculty member at Virginia Tech. My translational research laboratory is interested in identifying basic biological pathways and nutraceutical compounds that can be used to understand and treat genetic disease. While we have focused on adult obesity in general, our laboratory has also worked in the area of Prader-

Willi Syndrome. Since starting my independent laboratory in 1997, I have successfully trained 3 postdoctoral fellows, 25 graduate students, and 50+ undergraduate students. My current and past collaborations span the translational spectrum from basic cellular and molecular biologists, to neurobiologists, to economists and clinicians, and 6 of my undergraduate students have published with me based on their independent undergraduate projects. This book and the programs that are described in my two chapters represents one way for me to pay it forward for those who gave me a chance as an undergraduate, including Dr. John Pickrell (LITRI, deceased) and Dr. William Staaz (SUNY Fredonia) and Dr. Patricia Astry (SUNY Fredonia), for each of whom these chapters are dedicated.

Wendy Gibson graduated from Towson University with a degree in Elementary Education and a graduate degree in Reading. She currently teaches fifth grade in Baltimore County Public Schools and is a graduate student in Mathematics Education at Towson University.

Laura Gough is Professor and Chair of Biology at Towson University. She is a plant ecologist with expertise in wetland and tundra community and ecosystem ecology. At Towson she has been the PI of the HHMI IE project developed to create Course-Based Undergraduate Research Experiences (CUREs) and support faculty in learning about CURE pedagogy and inclusive teaching. She continues to explore ways in which her department and college can ensure STEM student success.

Rhett Herman received his B.S. in Physics and Chemistry from Wake Forest University, and his M.S. and PhD from Montana State University. His original research field was in semiclassical gravity, the uneasy alliance of General Relativity and Quantum Mechanics, studying the highly-curved spacetimes of charged black holes. At Radford University, his research has evolved into areas of near surface geophysics as a venue to involve undergraduates. He works closely with his students as they develop their own microcontroller-based sensor systems that they deploy in many locations, but especially on the arctic sea ice in Utqiagvik, Alaska, in an every-other year full academic year research experience.

Shawn M. Huston earned his bachelor's degree (2006) and PhD (2012) in physics from Ball State University and North Carolina State University respectively. After completing his doctorate he was a Visiting Research Professor at Appalachian State University before beginning his tenure in the physics department at Radford University in 2013. He is passionate about the success of his students.

Sarah A. Kennedy is an Associate Professor of Chemistry at Radford University, where she currently serves as the Program Director for both REALISE-Radford's HHMI Inclusive Excellence grant and the university-wide quality enhancement plan RISE-Realizing Inclusive Student Excellence. Both programs aim to increase students' academic success and belongingness through faculty development and student support. Her scientific areas of interest include chemical education, green chemistry, and biomolecular structure.

Jamie K. Lau is an aquatic ecologist at Radford University in Southwest Virginia and serves as the Chair of the Department of Biology. Jamie teaches courses in scientific communication, introductory and upper-level ecology courses, and senior seminar. Her main teaching goal is to facilitate the development of a student's science identity through authentic research experiences, both in the classroom and in her research program. She was part of the REALising Inclusive Science Excellence (REALISE) leadership team, active in the assessment of the program's outcomes.

Sandra Liss is an Assistant Professor of Physics and Director of Selu Observatory at Radford University in Radford, VA. She obtained her B.A. in Physics from Swarthmore College in 2011 and her M.S. and Ph.D in Astronomy from the University of Virginia (UVa) in 2014 and 2018, respectively. Her dissertation research at UVa focused on star formation in nearby interacting dwarf galaxies using large ground-based optical telescopes and space-based optical and infrared telescopes. In addition to teaching at Radford, Sandra is involved in implementing inclusive science education initiatives, developing a diverse and active astronomy research group, and restoring the telescope at Selu Observatory.

Sasha C. Marine is a Collegiate Assistant Professor with the Department of Biochemistry at Virginia Tech. Her primary teaching responsibilities are: general non-majors biochemistry (undergraduate), analysis of primary literature (undergraduate), and scientific communication (graduate). She is currently conducting research on the influence of topic order on student learning and success. Prior to joining the Department of Biochemistry in 2017, she was a postdoctoral associate with the University of Maryland and an adjunct instructor at the University of Maryland Eastern Shore.

Jeanne Mekolichick is Associate Provost of Associate Provost of Research, Faculty Success & Strategic Initiatives and Professor of Sociology at Radford University. She provides strategic leadership and direction for the research and creative scholarship enterprise, online education, faculty success, global education and engagement, experiential learning and strategic initiatives. During her tenure at Radford University, Dr. Mekolichick has served in a variety of leadership roles, winning awards for both teaching and leadership. Her work has been funded by public and private agencies for mission-central efforts including inclusive excellence initiatives, community-based research, and undergraduate research programming. Dr. Mekolichick earned her doctorate of philosophy in sociology also from Kent State University and is President Emerita of the Council on Undergraduate Research.

Patrice E. Moss is a Clare Boothe Luce Associate Professor of Biochemistry at Trinity Washington University (Trinity), where she also serves as the Biology Program chair. Dr. Moss's platform at Trinity has been focused on increasing women in color in STEM programs and careers by providing undergraduate research and experiential learning opportunities to students; thereby, broadening participation of minoritized groups in science disciplines. Dr. Moss is a proud graduate of the University of Maryland Eastern Shore (B.S., Biology) and Meharry Medical College (Ph.D., Biochemistry/Cancer Biology).

Lynn Nichols earned a B.A. in Mathematics and a master's in teaching from the University of Virginia. She is currently pursuing a Ph.D. in Instructional Technology at the College of Education at Towson University. Lynn is a PreK-12 Instructional Technology Coach and robotics, technology, and mathematics teacher. In addition to coaching faculty on traditional and maker-based pedagogies, Lynn creates STEM enrichment programming on robotics, laser cutting, 3D printing, digital citizenship, and computational thinking for PK-12 students. Her research interests include justice-centered learning and fostering inclusivity in makerspaces and using Arduinos to leverage computational thinking.

Deborah Pollio lives in the mountains of southwest Virginia. She is a retired instructor and advisor from Virginia Tech, where she worked in the engineering college as well as with pre-health students from the college of life sciences. Deborah has spent decades learning to connect with college-aged students. She has taught classes in a variety of styles and mediums, and is most interested in helping students engage with course material and connect back with faculty and with each other. Empowering students to succeed at the college level and to become part of the university culture has been a major focus of her work.

Amanda C. Raimer is a Postdoctoral Teaching Fellow and Program Manager within Radford University's HHMI IE program, Realising Inclusive Science Excellence (REALISE). She received her Ph.D. in Genetics and Molecular Biology from UNC Chapel Hill. At Radford, Amanda teaches in the Biology Department and engages undergraduates in course-based undergraduate research experiences (CUREs) as well as independent research through HHMI's SEA-PHAGES/SEA-GENES program. Through REALISE, Amanda mentors peer role models and develops inclusive programming for faculty and students with the goal of strengthening institutional practices that lead to success for all students.

Mia Ray is an Associate Professor of Biology at Trinity Washington University, specializing in the area of anatomy and physiology (A&P). She completed her undergraduate degree in Mathematics from Spelman College in Atlanta, GA and earned a doctorate in Anatomy from Howard University in Washington, DC. Dr. Ray has been teaching in the area of anatomy and physiology since 2008 and joined the faculty of Trinity Washington University in the fall of 2012.

Michael D. Schulz is an Assistant Professor in the Department of Chemistry at Virginia Tech, and a member of the Macromolecules Innovation Institute, the Center for Emerging, Zoonotic, and Arthropod-borne Pathogens, and the Virginia Tech Center for Drug Discovery. He received his Ph.D. in 2014 in organic and polymer chemistry and an M.S. in Pharmaceutical Science at the University of Florida under the supervision of Prof. Ken Wagener. After conducting research at the Max Planck Institute for Polymer Research as a Fulbright Scholar in the group of Prof. Klaus Müllen, he was a postdoctoral scholar in the group of Prof. Robert Grubbs at Caltech. He began his independent career at Virginia Tech in 2017.

Kristina Stefaniak is an Associate Professor in the Department of Chemistry at Radford University in Radford, VA. She obtained her B.S. in Chemistry from Temple University and her Ph.D. from Virginia Tech in 2017. Her research is focused on analytical chemistry and local environmental chemistry questions. Kristina works to develop inclusive curriculum in her general, analytical, and instrumental chemistry courses.

Brett Taylor, PhD, is a professor of physics and the department chair at Radford University. He has taught courses in General Physics I, General Physics II, Introductory Physics Seminar, Physics I, Physics II, Physics of Sound, Mathematical Methods in Physics, Numerical Methods in Physics, Electromagnetic Theory I, and Electromagnetic Theory II.

Kaitlin R. Wellens received her B.S. in Biology from Bates College and her PhD in Hominid Paleobiology from The George Washington University. She is currently a Clare Boothe Luce Assistant Professor of Biology at Trinity Washington University where her work focuses on both wild chimpanzee behavior and inclusive pedagogy in STEM.

This program is supported in part by a grant to Virginia Tech from the Howard Hughes Medical Institute through the Inclusive Excellence Grant.

REVIEWING, ADOPTING, OR ADAPTING THIS BOOK

If you are an instructor reviewing, adopting, or adapting this book, please help us understand a little more about your use by filling out this form: https://bit.ly/interest_fosteringcommunities. Instructors selecting the text are encouraged to fill out this form in order to stay up to date regarding collaborative development or research opportunities, errata, new volumes and editions, supplements and ancillaries, and newly issued print versions. Additional suggestions or feedback may be submitted via email at: publishing@vt.edu.

ADDITIONAL RESOURCES

These following resources for *Fostering Communities of Transformation in STEM Higher Education* are available at https://doi.org/10.21061/fosteringcommunities

- Free and openly licensed downloadable PDF documents
- Free and openly licensed EPUB
- A link to purchase a print copy (sold at cost)

WHAT IS AN OPEN TEXTBOOK?

Open textbooks are complete textbooks that have been funded, published, and licensed to be freely used, adapted, and distributed. As a particular type of open educational resources (OER), this open textbook is intended to provide authoritative, accurate, and introductory level subject content at no cost, to anyone including those who utilize screen reader technology to read and those who cannot afford traditional textbooks. This book is licensed with a Creative Commons Attribution 4.0 International license, which allows it to be adapted, remixed, and shared with attribution. Professors and others may be interested in localizing, rearranging, or adapting content, or in transforming the content into other formats which meet the goal of better addressing student learning needs, and/or making use of various teaching methods.

Open textbooks are available in a variety of disciplines via the Open Textbook Library: http://open.umn.edu/opentextbooks.

CHAPTER 1.

PREPARING STEM TEACHERS TO BE CHANGE MAKERS

KIMBERLY CORUM AND LYNN NICHOLS

ABSTRACT

In this chapter, we will share an instructional technology graduate course designed to introduce teachers to emerging technologies commonly found in makerspaces and how making can be used to understand and address complex social justice problems. Analysis of students' submitted coursework and their post-course reflections revealed that exposure to social justice mathematical making lessons supported a shift in their beliefs about incorporating both technology and social justice contexts into their classrooms.

Broadly defined, a makerspace is a physical space with the necessary tools and materials to encourage creative design (Cavalcanti, 2013). It is estimated that there are more than two thousand makerspaces across the United States (Nation of Makers, 2022) and these spaces are becoming increasingly present in K-12 schools (Peppler & Bender, 2013). Makerspaces have the potential to transform STEM education because they provide students with opportunities to investigate authentic mathematical and scientific questions that arise organically through project design and development (e.g., Blikstein, 2013; Martin, 2015). As further explained by Blikstein (2013), "Abstract ideas such as friction and momentum become meaningful and concrete when they are needed to accomplish a task within a project" (p. 18). Makerspaces can also be leveraged to promote equity in science, technology, engineering, and mathematics (STEM) education, particularly when coupled with justice-centered pedagogies (e.g., Barton, Tan, & Greenberg, 2017; Vossoughi, Hooper, & Escudé, 2016). In order for makerspaces to be realized to their full potential, teachers need professional development focused on integrating emerging technologies with teaching for social justice.

Otherwise, the warning issued by Seymour Papert nearly thirty years ago will still hold true: "The phrase 'technology and education' usually means inventing new gadgets to teach the same old stuff in a thinly disguised version of the same old way" (Papert, 1980, p. 1).

INCLUSIVE MAKERPACK

Effectively integrating new technologies into classroom practice requires teachers to have a specialized knowledge known as technological pedagogical content knowledge (TPACK). Building Shulman's (1986) pedagogical content knowledge framework, TPACK describes the integration of technological expertise with an understanding of how technology can support content-specific learning (Koehler & Mishra, 2009). However, the TPACK framework attends to technological knowledge more broadly and does not consider practices typically associated with making (e.g., design thinking, problem solving). The MakerPACK framework (Figure 1.1; Corum, Spitzer, Nichols, & Frank, 2020) builds on the TPACK framework by considering the novel features of emerging technologies commonly found in makerspaces (e.g., digital fabrication, coding, robotics, microcontrollers) and how these technologies work in conjunction with design thinking.

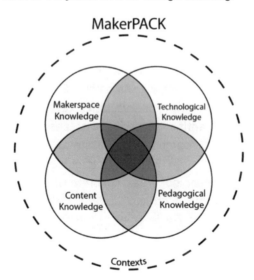

Figure 1.1: MakerPACK Framework

Given the potential of makerspaces to increase student engagement and participation in STEM fields, it is paramount that these spaces are inclusive for all students and that teachers are equipped to meet this need. Unfortunately, not all students feel as if they belong in makerspaces. A study of makerspaces across the United States that offered K-12 educational programming revealed a pervasive gender bias in these spaces. For example, researchers found that the identity markers most often used by instructors to describe male students included "geeks," "builders," "designers," and "engineers," whereas the identity markers most often used to describe female students included "girls" and "helpers" (Kim, Edouard, Alderfer, & Smith, 2018). The fact that these identity markers were used by educators in these spaces is particularly problematic. For an in-depth discussion of Black, Indigenous, and people of color (BIPOC) students' experiences in makerspaces, please refer to the chapter, "Strategies for Creating and Sustaining Inclusive Makerspaces" later in this book. In response to these concerns, we offer an extension of the MakerPACK framework that integrates teacher knowledge of the needs of students who identify as BIPOC (Figure 1.2; Nichols & Corum, 2023).

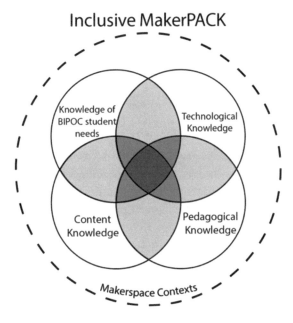

Figure 1.2: Inclusive MakerPACK Framework

While there is limited research regarding how to best support teachers in utilizing makerspaces in their classroom instruction, there is evidence that meaningful professional development is critical in supporting teachers' development of TPACK, and their ability to integrate technology into the classroom requires meaningful professional development (e.g., Wenglinksy, 1998; Koehler & Mishra, 2005; Bos, 2011). Guided by the Inclusive MakerPACK framework, we considered how to prepare teachers to integrate making and makerspaces to support mathematics teaching in a way that centers equity and access. In the following sections, we will share our approach and our findings.

COURSE DESIGN AND DEVELOPMENT

Using Makerspace Technology in School Mathematics is an instructional technology course that is offered to students enrolled in the Mathematics Education M.S. degree program at Towson University. The course is organized around five technologies commonly found in makerspaces: paper circuitry, three-dimensional (3D) design and fabrication, coding, robotics, and Arduino microcontrollers (Corum et al., 2020). The course was first offered in Fall 2019 and since then, the course has been offered an additional five times across three different semesters (Spring 2020, Fall 2021, Fall 2022).

Prior to the Fall 2022 offerings, the course content was redesigned to highlight how both making and mathematics could be used to address social justice issues. Five social justice-centered mathematical making (SJMM) lessons were developed to complement the five central technologies of the course (Nichols & Corum, 2023) and to demonstrate authentic applications of making to understand and solve real-world problems. These lessons are described below.

SEA LEVEL CHANGE AND PAPER MATHEMATICS

Adapted from Archey's (2019) "Sea Level Change and Function Composition" activity, this lesson explores the impact of sea level change on our planet. After analyzing National Oceanic and Atmospheric Administration (NOAA) interactive data to visualize the current and future impacts of sea level change, learners create paper models (Figure 1.3) to demonstrate the relationship between changing sea levels and remaining land. In the model shown in Figure 1.3, the triangular prism represents land, and the overlaid paper (which can be manipulated) represents sea level.

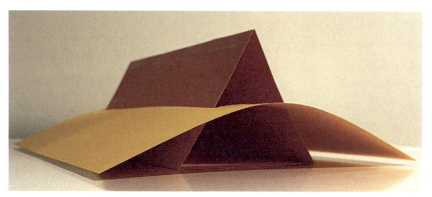

Figure 1.3: Paper model to represent sea level change

MONUMENT STORIES AND 3D PRINTING

This lesson investigates the Daughters of the Confederacy's influence on our nation's monuments. After analyzing findings of the National Monument Lab audit, learners use 3D design to create their own monuments to celebrate unsung heroes in our nation's history (Figure 1.4) and apply proportional reasoning to consider a full-scale model of their designs.

Stand your ground by having a seat

Figure 1.4: Student-designed monument to honor Claudette Colvin

MODELING FOOD ACCESSIBILITY WITH PROGRAMMING IN SCRATCH

This lesson examines the issue of food security in Baltimore. After deciding on their preferred location for a grocery store in a neighborhood that has been classified as a healthy food priority area, learners use Scratch (a block-based coding language) to model walking and driving conditions to evaluate the accessibility of their proposed grocery store location (Figure 1.5).

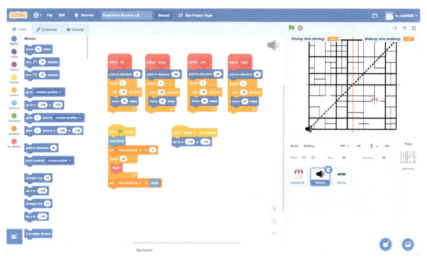

Figure 1.5: Scratch program shell

LEGO PROSTHETICS AND ROBOTICS

This lesson considers the challenges of accessing and receiving prosthetic limbs. After seeing examples of prosthetic limbs constructed from LEGO robotics pieces, learners prototype their own prosthetics using LEGO pieces and determine the torque needed to stall the motor of their builds (Figure 1.6).

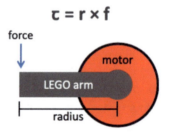

Figure 1.6: Model used to determine stalled torque of LEGO prosthetic arm

WATER CONSERVATION WITH ARDUINO

This lesson studies the issue of water loss in farming. After learning about vertical farming as an alternative farming method that minimizes water loss, learners program an Arduino with a moisture sensor to create a device that can monitor soil water levels (Figure 1.7).

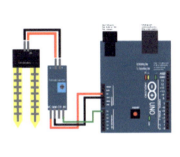

```
int sensor_pin = A0;
int output_value;

void setup() {
Serial.begin(9600);
Serial.println("reading from the sensor…");
delay(2000);
}

void loop() {
output_value = analogRead(sensor_pin);
output_value =
map(output_value,550,0,0,100);
Serial.print("Moisture: ");
Serial.print(output_value);
Serial.println("%");
delay(1000);
}
```

Figure 1.7: Wiring diagram and code for building a moisture sensor

---

DATA COLLECTION AND ANALYSIS

To understand how participating in the redesigned *Using Makerspace Technology in School Mathematics* course supported teachers' development of an Inclusive MakerPACK, we developed a post-course survey (adapted from Enterline et al., 2008, Appendix A) that asked participants to reflect on how their course experiences impacted their attitudes toward incorporating justice-centered mathematics and/or mathematical making in their own classroom practice. Our participants included graduate students enrolled in one of the two sections of the course offered during the Fall 2022 semester. This section included graduate students in the Mathematics Education M.S. degree program who are not part of a school district partnership which allowed for greater diversity in participants' teaching contexts. Of the fourteen students enrolled in this section of the course, seven students submitted survey responses.

During the first round of data analysis, we established a list of *a priori* parent codes (e.g., impactful activities, new beliefs, prior beliefs, obstacles, solutions) and then recorded child nodes (e.g., 3D design, coding, importance of social justice, limited experience) associated with each parent code during the initial pass (Merriam & Tisdell, 2015; Saldaña, 2021). Prior to the second round of data analysis, the parent codes were refined or redefined based on themes observed during the first round of data analysis. During the second round of data analysis, the existing child nodes were redefined, regrouped, and refined, and additional child nodes were identified based on patterns observed in the data. We then discussed our findings before the third and final round of data analysis to confirm that the most salient themes in the data were identified (see Table 1.1).

FINDINGS

Analysis of survey results revealed that participating in *Using Makerspace Technology in School Mathematics* course impacted graduate students' beliefs about incorporating social justice-centered mathematical making activities in their own classrooms. All participants who submitted the survey indicated that at least one of the SJMMs from the course truly resonated with them and that they intended to apply what they learned from the course in their own practice. The main themes that emerged from students' post-course surveys are summarized in Table 1.1. These themes were consistently present across all participants' survey responses and often referenced multiple times.

Table 1.1: Main themes from post-course surveys (n = frequency count)

Theme	Explanation	Example
Prior Beliefs (n = 14)	Participants shared that they had not previously integrated making and/or social justice into their own practice and attributed this to their own limited experience with these concepts.	"I have never seen SJ [social justice] integrated within a math classroom and did not think it was possible to do so."
Impactful Activities (n = 15)	Participants shared that the most impactful SJMMs were those that centered promoting equity and those that made technologies that previously seemed too advanced given their current expertise more accessible.	"...being more aware of marginalized populations and making mathematical connections for the classroom was refreshing." "[Robotics] seems intimidating but is actually very accessible and user friendly"
New Beliefs (n = 35)	As a result of their course experiences, participants shared that SJMMs helped them realize the value of integrating social justice contexts into their classrooms and the benefits of using technology to make content more accessible for all learners.	"I enjoyed learning how SJ [social justice] can be incorporated into the math classroom and plan on incorporating SJ into my classroom and instruction." "...using the Ozobotos may help some of my students on the spectrum engage with math using colors and shapes..."
Obstacles and Solutions (n = 9; n =17)	Participants anticipated pushback from students and administration related to the perceived political nature of SJMMs and concerns with maintaining fidelity with assigned curricula/pacing guides as potential obstacles to implementing SJMMs in their own contexts, but found that purposeful professional development related to this approach to teaching and learning could provide a solution.	"Students must grapple with certain data that is intentionally misleading, which causes cognitive dissonance when they already had false beliefs from similar data." "Trying to convince planning group with the curriculum time constraints."

Graduate students' coursework provides additional evidence of the impact of the course on their professional development, particularly related to applying their experiences to support their own classroom instruction. For each technology explored in the course, graduate students were asked to design and develop their own lessons that incorporated the specified technology. It was not a requirement for students to also incorporate a social justice context in these lessons. However, several teachers chose to develop SJMMs.

> These lessons included:
>
> - Addressing inequitable access to resources and using triangle paper constructions to identify the equitable placement of cellular towers and grocery stores;
> - Understanding the experiences of environmental refugees impacted by climate change and using quadratics to model land loss of Tangier Island;
> - Investigating redistricting through a socio-economic and racial lens and using geometry and proportional reasoning to develop a scale model of a school building to meet the evolving demands resulting from redistricting in Montgomery County, Maryland

---

CONCLUSION

Analysis of post-course reflections and submitted coursework revealed that exposure to social justice mathematical making (SJMM) lessons in the *Using Makerspace Technology in School Mathematics* course supported a shift in graduate students' beliefs about incorporating both technology and social justice contexts into their classrooms. Participating in the SJMM lessons allowed our graduate students to experience firsthand how mathematics and technology can be used to understand and address social injustices. For many of our graduate students, this was a transformative experience as it was the first time they had considered incorporating social justice into a mathematics classroom. While we were able to redesign our graduate course to center social justice, we recognize that others may not be able to do the same. However, teacher educators can still find opportunities to incorporate social justice topics into their content and methods courses. Modeling how to incorporate issues related to diversity, equity, inclusion, and justice into classroom practice can be a valuable first step when preparing teachers to be change makers.

Acknowledgments

This project is based upon work supported by the National Science Foundation under Grant No. 2243461. Any opinions, findings, and conclusions or recommendations expressed in this material are those of the author(s) and do not necessarily reflect the view of the National Science Foundation.

APPENDIX A

Using Makerspace Technology in School Mathematics Post-Course Survey

1a) Before this course, what were your beliefs and/or prior experiences with justice-centered mathematics?

1b) Share which lesson and/or class experience had the greatest impact on your beliefs about incorporating justice-centered lessons in your mathematics classroom? Why were these impactful? The social justice mathematics lessons we explored this semester included Sea Level Change (Paper Engineering), Monument to an Unsung Hero (3D Design), Food from Scratch (Coding), LEGO Prosthetics (Robotics), and Water Conservation (Arduino).

1c) Discuss how your thinking about justice-centered mathematics has changed since participating in the five social justice mathematics lessons. What do you believe now and why did your thinking change?

2a) What topic did you choose? Why did you choose this topic?

2b) How might you include this topic in your curriculum?

2c) Discuss what hurdles you anticipate encountering when incorporating this justice-centered lesson into your curriculum.

2d) What are some possible solutions to overcome these hurdles?

3a) Before this course, what were your beliefs and/or prior experiences with mathematical making or making in general?

3b) Share which lesson and/or class experience had the greatest impact on your beliefs about mathematical making in the classroom. Why were these impactful? Examples of the mathematical making lessons we explored this semester include finding the volume of the origami cube, designing the equal probability dice, writing a Python program to draw a polygon, and collecting data with Spheros.

3c) Discuss how your thinking has changed since participating in the mathematical making lessons. What do you believe now and why did your thinking change?

(Optional) Do you have any additional comments?

REFERENCES

Barton, A. C., Tan, E., & Greenberg, D. (2017). The makerspace movement: Sites of possibilities for equitable opportunities to engage underrepresented youth in STEM. *Teachers College Record*, *119*(6), 1-44.

Blikstein, P. (2013). Digital fabrication and 'making' in education: The democratization of invention. In J. Walter-Herrmann & C. Büching

(Eds.), *FabLabs: Of Machines, Makers, and Inventors*. Bielefeld, Germany: Transcript Publishers.

Bos, B. (2011). Professional development for elementary teachers using TPACK. *Contemporary Issues in Technology and Teacher Education, 11*(2), 167-183.

Cavalcanti, G. (2013). Is it a Hackerspace, Makerspace, TechShop, or FabLab? Makezine. https://makezine.com/2013/05/22/the-difference-between-hackerspaces-makerspaces-techshops-and-fablabs/

Corum, K., Spitzer, S., Nichols, L., & Frank, K. (2020). Developing TPACK for makerspaces to support mathematics teaching and learning. *Mathematics Education Across Cultures: Proceedings of the 42nd Meeting of the North American Chapter of the International Group for the Psychology of Mathematics Education*, 2193–2197. https://doi.org/10.51272/pmena.42.2020-374

Enterline, S., Cochran-Smith, M., Ludlow, L. H., & Mitescu, E. (2008). Learning to teach for social justice: Measuring change in the beliefs of teacher candidates. *The New Educator, 4*(4), 267–290. https://doi.org/10.1080/15476880802430361

Kim, Y. E., Edouard, K., Alderfer, K., & Smith, B. K. (2018). *Making culture: A national study of educational makerspaces*. Drexel University ExCITe Center. https://drexel.edu/~/media/Files/excite/making-culture-full-report.ashx?la=en

Koehler, M., & Mishra, P. (2005). Teachers learning technology by design. *Journal of Computing in Teacher Education, 21*(3), 94-102.

Koehler, M., & Mishra, P. (2009). What is technological pedagogical content knowledge (TPACK)? *Contemporary Issues in Technology and Teacher Education, 9*(1), 60–70.

Martin, B. (2015). Successful implementation of TPACK in teacher preparation programs. *International Journal on Integrating Technology in Education, 4*(1), 17-26.

Merriam, S. B., & Tisdell, E. J. (2015). *Qualitative research: A guide to design and implementation* (Fourth edition). John Wiley & Sons.

Nation of Makers (2022). *Making the Case.* https://www.nationofmakers.us/makingthecase.

Nichols, L., & Corum, K. (2023). Increasing teacher commitment to justice-centered mathematics through maker-enhanced social justice activities. In E. Langran, P. Christensen, & J. Sanson (Eds.), *Proceedings of Society for Information Technology & Teacher Education International Conference* (pp. 411-416). New Orleans: Association for the Advancement of Computing in Education (AACE). https://www.learntechlib.org/p/221891/

Papert, S. (1980). *Mindstorms: Children, computers, and powerful ideas.* New York, NY: Basic Books, Inc.

Peppler, K., & Bender, S. (2013). Maker movement spreads innovation one project at a time. *Phi Delta Kappan, 95*(3), 22–27. https://doi.org/10.1177/003172171309500306

Saldaña, J. (2021). *The coding manual for qualitative researchers* (4th ed). SAGE Publishing.

Shulman, L. S. (1986). Those who understand: Knowledge growth in teaching. *Educational Researcher, 15*(2), 4-14.

Vossoughi, S., Hooper, P. K., & Escudé, M. (2016). Making through the lens of culture and power: Toward transformative visions for educational equity. *Harvard Educational Review, 86*(2), 206-232.

Wenglinsky, H. (1998). *Does it compute? The relationship between educational technology and student achievement in mathematics.* Educational Testing Service. https://www.ets.org/Media/Research/pdf/PICTECHNOLOG.pdf

---

This program is supported in part by a grant to Virginia Tech from the Howard Hughes Medical Institute through the Inclusive Excellence Grant.

CHAPTER 2.

RESEQUENCING THE CHEMISTRY CURRICULUM TO RETAIN CHEMISTRY MAJORS

Optimizing connections between general and organic chemistry
MICHAEL D. SHULZ

ABSTRACT

Why do students become chemistry majors? Why do they stay chemistry majors? The answers to these questions vary by student, but themes do emerge. Often, an inspirational teacher in high school plants an idea in a student's mind that they could become a scientist, perhaps specifically a chemist. Once that student arrives in college, however, that idea too often withers, and the student leaves for another discipline. What can we, the Chemistry Department faculty, do to nurture that burgeoning interest in our field? Again, the answers vary. Often, the answers are different for students from underrepresented groups. At Virginia Tech, a larger R1 institution, we decided to try an experiment: We would give students as wide a perspective on chemistry as possible by moving our Organic Chemistry 1 course into the freshman year of our curriculum. This change would expose students to both the quantitative aspects of chemistry typically presented in the General Chemistry curriculum, as well as the more qualitative thinking embedded in Organic chemistry.

INTRODUCTION

OUR HYPOTHESIS

Our faculty generally fall into two camps: those who like organic chemistry and those who don't. Students are arguably the same. Granted, this distinction isn't hard and fast—many of my colleagues like all areas of chemistry, and most excelled in every chemistry course they took. But still, this distinction remains perceptible. Countless students and faculty alike can share stories of almost leaving chemistry either until they reached their first Organic Chemistry class and discovered a new passion for the discipline, or they reached their first Organic Chemistry class and couldn't wait to be done with it. To reveal my own allegiance, I fall into the former camp, having had only a vague interest in general chemistry but thoroughly enjoying my organic chemistry courses and research in a synthetic chemistry lab.

The existence of this dichotomy surprises few chemists. In fact, I've occasionally heard senior students advise freshmen to "wait until you've had Organic" before making any definitive judgement on pursuing a career as a chemist. Whether this advice is encouraging or foreboding depends on the context.

Like many institutions, we followed a course structure that featured a year of General Chemistry—introducing foundational concepts, but largely as preparation for analytical and physical chemistry courses—followed by a year of Organic Chemistry: General Chemistry 1, General Chemistry 2, Organic Chemistry 1, Organic Chemistry 2. Many students, however, do not remain chemistry majors long enough to experience Organic Chemistry. Many students leave our major after the first year (or even the first semester). The causes of these departures are multifaceted, but we hypothesized that perhaps we are losing some students who would have discovered that they liked organic chemistry, if they had the opportunity to take it as a freshman. We wanted to make an effort to retain these students.

CONSIDERING A CHANGE

We began exploring a reorganization of our Chemistry Majors curriculum, specifically moving Organic Chemistry into the freshman year. The new course order would then be Gen Chem 1, Organic 1, Organic 2, Gen Chem 2, an order that is sometimes referred to as "1-2-2-1". Several advantages to this curriculum structure soon became obvious:

Advantages to this Curriculum Structure:

- A broader spectrum of chemical topics could be covered in the freshman year, thereby exposing students to some of the less quantitative organic chemistry concepts that many students find more enjoyable.

- Gen Chem 2 would immediately precede Analytical and Physical Chemistry, which would ideally enable the students to have key concepts from Gen Chem fresh in their minds as they start learning the more advanced topics. In the original curriculum, the entire sophomore year would elapse between Gen Chem and Analytical chemistry, producing a "great forgetting" among the students. As Organic is not a prerequisite for either Physical or Analytical Chemistry, this change would potentially group similar material together more effectively.

- Gen Chem 2 could potentially function as a sophomore-level course, perhaps better preparing students for the upper-division chemistry classes.

This curriculum reorganization offered us the opportunity to address at least one feature of the curriculum that seemed suboptimal, but persisted partly due to institutional inertia: the relationship between the Organic Chemistry Lecture and Organic Chemistry Lab.

Our Organic Chemistry Lecture and Organic Chemistry Lab courses are separate entities, completely independent of each other and typically taken off-sequence (i.e., Organic Lab 1 is typically taken in the second-semester sophomore year, during the Organic 2 Lecture course; Organic Lab 2 is then taken as a junior with no accompanying Organic Chemistry Lecture class). Thus, reorganizing the curriculum offered us the opportunity to align Organic Chemistry Lecture and Lab so that they could mutually reinforce themes and concepts. Our students had long requested this integration, and here was our opportunity to do so.

Another factor also featured prominently in our thinking: We wanted to better support our students in achieving the math readiness needed to succeed in chemistry. More and more students needed additional math courses before taking calculus, which is our assumed starting point for a freshman chemistry major. Additionally, many Gen Chem students could grasp the chemistry concepts being taught, but sometimes struggled with the math component on exams. A fascinating discussion may be had on the extent to which General Chemistry exam questions should require mathematical manipulations, but the bottom line was nonetheless clear: Students needed more time in the curriculum to reach the math proficiency required to succeed in chemistry. Pivoting to a more organic chemistry-focused freshman year would allow that space for students to build their math skills, thereby better setting them up for success in the more quantitative elements of Gen Chem 2.

Having determined that some benefits might result from such a curriculum, we started gathering perspectives on what such a curriculum revision might entail, what challenges might arise, and whether the potential advantages were worth the cost.

––––––––––

GATHERING PERSPECTIVES

Who else has done something like this? The answer, it turns out, is complicated. Several institutions employ the 1-2-2-1 curriculum structure, but each used it for different reasons, and each had different strengths and challenges. Some had used this course sequence since time immemorial, and it was accepted by their students as "normal". In our research, we didn't find an institution that had tried such a course resequencing for our specific reasons. Nevertheless, we had some conversations.

Early on, we talked with a chemistry faculty member at a peer institution that employed the 1-2-2-1 course structure. Their reasons for adopting this approach were quite different from ours, however. First, a large percentage of their students had high scores on AP Chemistry exams, which enabled them to bypass a portion of the Gen Chem curriculum. The 1-2-2-1 course structure solved certain scheduling issues. Second, most of their chemistry students were pre-med, and they wanted to get them into Organic Chemistry earlier with the belief that the Organic content is perceived to be more medicine-relevant by the pre-med students. Additionally, this institution did not have separate Gen Chem and Organic classes for their chemistry majors. This set-up is partly because their students do not declare majors immediately upon matriculation, so everyone takes the same sequence of chemistry classes, which are designed to serve all majors.

Our situation was different in several respects. While many of our incoming students arrived with AP credit, including several with strong scores on the AP Chemistry exam, we generally advise everyone to take our regular Gen Chem sequence. We reasoned that this approach would enable us to establish a common knowledge baseline across all students. However, we also observed a strong predictive quality to students' previous experience with AP Chemistry—students with AP Chemistry credit generally do well in our Gen Chem class, while students without such AP experience often struggle. An argument can be made that our approach was producing classes populated with students with wildly different levels of preparation, thereby setting some students up for failure despite our best efforts to help everyone succeed. We were concerned that continuing this approach might perpetuate disparities.

Pre-med students, of course, take General and Organic Chemistry, but few of them are chemistry majors at Virginia Tech. However, while our pre-med student numbers are low (perhaps unusually low), we typically have many students interested in engineering. Virginia Tech has long had a strong engineering program, and often students are interested in chemistry, but with an eye toward engineering applications. These students plan to pursue the engineering-adjacent aspects of chemistry, which are sometimes not obviously connected with organic chemistry. As we contemplated the curriculum reordering, we often considered potential effects on these groups of students.

Beyond our colleagues at other institutions, we also had conversations with our current chemistry majors. These conversations confirmed many of our assumptions about the perception of our program—that it was incredibly challenging, that pre-med students opted for perceived "easier" majors, that the underlying themes of Gen Chem were sometimes lost on the students, which produced a disjointed feel to the class. We asked if moving Organic 1 into the first year of the curriculum would be useful, and the responses seemed to follow the same dichotomy we had among the faculty: The students who loved organic thought it was a good idea, the students who hated organic wondered why we would ever consider doing such a thing.

ANALYZING OUR CURRICULA

Recognizing that our curriculum may not be attracting key groups of students who could benefit from our program, we had recently introduced two new chemistry majors: Medicinal Chemistry and Polymer Chemistry. The thought was that Medicinal Chemistry could attract the chemistry-inclined pre-med students, while the Polymer Chemistry degree might attract those with an interest in engineering. Consequently, when we started thinking about resequencing our curriculum, we had four majors to consider, each with its own requirements: Chemistry (BS), Chemistry (BA), Medicinal Chemistry (BS), and Polymer Chemistry (BS). Any curriculum changes would affect only the relatively small number of chemistry majors; the non-majors curriculum would remain unchanged because we couldn't accommodate a large-scale course-sequence switch due to limited lab capacity. But first we had to decide what part of Gen Chem was absolutely required before students start learning Organic.

We started by asking, "What are students currently taught in Gen Chem?" We got a list, which we analyzed with regard to organic chemistry relevance. Some judgements were easy: Structure and bonding are critical concepts for organic chemistry, radiochemistry probably less so. Other content produced mixed opinions. In general we thought it was desirable to make few large changes so as to streamline any potential resequencing; however, we struggled in making a distinction between topics that "are relevant to" organic chemistry at the sophomore level versus topics that are "essential to understand" organic chemistry at the sophomore level.

This discussion was complicated by the fact that closely related concepts often appear quite different to students in the context of organic chemistry versus general chemistry. Acids and bases are a prime example: Much of organic chemistry can be described in terms of acid-base reactions (e.g., Lewis acid/base in Gen Chem becomes electrophile/nucleophile in Organic), but students see the aqueous inorganic acids/bases of Gen Chem as a distinctly separate topic. Still, is half a lecture on polyprotic acids foundational to understanding organic chemistry? On the other hand, having Gen Chem 2 after the organic sequence offers untapped possibilities to use *organic* examples when teaching concepts, possibly reducing the compartmentalization of knowledge that many students seem to develop as they treat each class as unrelated to any others.

Beyond the specific ordering of the course content, a general consensus emerged that we were trying to teach too much—the students would be better served by presenting fewer topics but in greater depth. Said differently, we would prefer to have students master the truly foundational content rather than have a vague awareness of a greater breadth of topics. Of course, which content is "truly foundational" is often in the eye of the beholder.

Ultimately, we decided that the specific course content will need to be coordinated by the actual course instructors. With judicious but relatively minor adjustments to Gen Chem 1 and Organic 1, we thought that these two courses could function as the freshman chemistry majors introductory curriculum. Since the change was limited to the chemistry majors curriculum only (i.e., other majors would continue to take the regular non-majors course sequence), each course had only a single instructor, which we anticipated would facilitate coordination. Despite some well-founded reservations, we decided to reorganize our curriculum.

TAKING THE PLUNGE

Change of any kind inevitably foments complaints, many legitimate but some less so. We knew that an adjustment of this scale would cause problems of various kinds and would be compared unfavorably to the way things used to be. Consequently, we planned to maintain the new curriculum structure for at least six years, so as to erase the institutional memory (in the students at least) that there was once a different curriculum. To make an apples-to-apples comparison, this new course sequence had to be perceived as "normal."

Of course, the first year would be the most challenging. We would have freshmen on the new sequence and sophomores on the old sequence. Simultaneously, we would have to teach both sequences, effectively doubling our teaching load, particularly in the transition year. Two additional factors introduced further complications: COVID-19 and significant changes in our organic faculty.

We began discussing these curriculum changes in 2019, not knowing that in a few months our curriculum, teaching experiences, and lives would radically change. As COVID-19 triggered radical changes in higher education (and everything else), we continued our discussions and planning, knowing that our ability to make clear before-and-after comparisons was now compromised. The new curriculum took effect in

Fall 2022, arguably a time when "normal" had returned. My own opinion, however, is that the ripple effects of the pandemic will continue for years, both in higher education and elsewhere. Deconvoluting the effects of our curriculum reorganization from the ongoing effects of COVID-19 may not be possible.

Like all departments, we have faculty joining and leaving the department most years. As chance would have it, our curriculum reorganization coincided with relatively significant changes in our organic faculty, with multiple retirements, departures, and new hires happening within the span of a couple years. Consequently, much of the increased teaching load associated with the curriculum transition fell on faculty who weren't a part of the department when these discussions first took place. Ideally, the same faculty that taught Gen Chem and Organic before the curriculum resequencing would also teach it after the resequencing to enable more accurate comparison. Unfortunately, that approach wasn't possible.

CHALLENGES, FORESEEN AND UNFORESEEN

While the official curriculum change started in Fall 2022, the real changes didn't start taking place until Spring 2023. At that point, we had freshmen taking Organic 1, sophomores taking Organic 2, and both taking Organic Lab. Much of our focus in the planning stage had been on the Gen Chem content and how it prepared students for Organic. Comparatively less attention was paid to the connections between Organic Lecture and Organic Lab.

We had hoped that this curriculum revision would enable better integration between the lecture and lab content, a change that students had requested for years. But as the Spring semester began, we quickly realized that the lecture-lab offset had a key advantage: Students starting the lab had already had a semester of organic chemistry. Consequently, a considerable amount of conceptual knowledge was already in place on Day 1 of the lab, and the course was designed with much of this knowledge assumed. We were now in a situation where the lab course was filled with double the usual number of students, comprising both freshmen taking organic for the first time and sophomores who were already in the second semester of the organic chemistry sequence.

While this challenge was not entirely unforeseen—we had discussed at length the necessity of revising the lab course to account for the fact that it would now be a freshman class—it was perhaps under appreciated. Consequently, the lab and lecture were not as coordinated as we had hoped. Lab content that built on Organic 1 Lecture content had to be taught in greater detail to the freshmen who were seeing it for the first time. Lab-related questions came up in Organic 1 Lecture that required knowledge of Organic 2 content. Freshman students felt underprepared at times for the lab content relative to the sophomores in the course.

On top of these challenges, the current freshmen now look forward to a sophomore year that looks quite different from what the current sophomores are experiencing. In the past, different classes (e.g., freshmen and sophomores) had relatively limited interactions, but this year they're all in the same lab, so the students talk. The rumored consensus is that the sophomores are glad they went through their classes on the old sequence, as now the freshmen are anticipating a sophomore year that looks particularly daunting. Many faculty generally agree that it is daunting, but the sophomore year was always daunting in our curriculum. We don't think the difficulty changed (it's arguably easier), but that perception is not necessarily shared by our students.

From the beginning, we tried to get broad buy-in for this plan at each stage of the process. For example, we discussed the proposed changes at meetings of the full faculty and the organic division specifically. We got the opinions of current students (and discarded several curriculum possibilities based on their feedback). Nevertheless, these efforts were imperfect, and as the curriculum changes were implemented, we heard complaints of poor planning leading to foreseeable problems. On the whole, though, major disasters have so far been averted—a particularly noteworthy accomplishment given the double enrollment in the lab course. Planning, implementation, and coordination across multiple courses continues to develop as we progress through this curriculum revision, but discovering the full impact of these changes will require a few years of data and perspective. Whether we will realize our stated goals remains to be seen.

LESSONS LEARNED AND FUTURE OUTLOOK

So, what have we learned? Perhaps the biggest lesson has been the importance of involving the people actually teaching the courses throughout the planning process from the very beginning. While various factors made that approach impossible in our case, we can see in retrospect that planning and implementing must go hand in hand, and that imperative is greatly facilitated when the same group of people are involved in both.

Additionally, we perhaps relied too much on informal coordination between the various faculty members involved in teaching these classes. Our faculty is highly collegial and collaborative, and such informal coordination is often sufficient (or even preferred). In this case, however, we may have benefited from also including a more formalized coordination effort, even if it only served to provide a designated space and time for discussing and planning. With the benefit of recent experience, we are now evaluating how we can best redesign and coordinate the courses (particularly Organic Lab) to better teach our students. I expect next year (and later years) will be better, as we continually refine our curriculum.

This chapter tells a story without an end. The value of this approach, however, is that our thinking can be captured without the bias of hindsight. At the time of this writing, we're still in the midst of the curriculum transition. While many of the challenges inherent in reorganizing a curriculum have already manifested, our hoped-for gains are still in the future. Consequently, the final success or failure of this experiment is unclear. Time will tell.

This program is supported in part by a grant to Virginia Tech from the Howard Hughes Medical Institute through the Inclusive Excellence Grant.

CHAPTER 3.

DELAYED ENROLLMENT IN GENERAL CHEMISTRY RECITATION

PATRICIA AMATEIS

ABSTRACT

Due to the large size (three hundred students per section) of General Chemistry classes at Virginia Tech, these freshman-level classes have generally been taught using a traditional lecture format because it is difficult to incorporate meaningful active learning in the class meetings. Many students do not respond well to traditional large lecture settings; to reach these students, we want to use small group recitations to provide opportunities for one-on-one instructor-student interactions, active learning, and a community learning environment. The large number of students taking General Chemistry in the Fall semesters prohibits us from providing small group recitations for all students. The challenge is identifying the students who most need the extra help provided in the recitations. To do this, we used the grades on the first exam in the class to identify a target group to receive an invitation for enrollment in a recitation class in the sixth week of the semester. Undergraduate science majors, supervised by a faculty member, were the recitation instructors. Analysis of grade data revealed marked improvement for the students who enrolled in the recitations. Grade data and student opinions about the recitation in the Fall 2020 and Fall 2021 semesters will be discussed.

INTRODUCTION

General Chemistry (CHEM 1035) at Virginia Tech is a large service course that presents challenges to both students and instructors. First, the students in CHEM 1035 represent a wide variety of majors and a wide range of academic backgrounds and math preparedness. Approximately half of our CHEM 1035 students are General Engineering majors and the other half have majors in Biological Sciences, Biochemistry, Animal and Poultry Sciences, Dairy Science, Human Nutrition, Food, and Exercise, Geoscience, Neuroscience, Horticulture, Environmental Science, Forestry, Wildlife, Nanoscience, and Natural Resources. Some students have strong backgrounds in calculus, while others begin their college careers in college algebra.

Secondly, class sections are large, with about 300 students in each of several sections for a total of about 2500 students each semester. Due to the large class size, a traditional lecture style has been implemented. Many students do not learn well in this type of teaching environment.

It has been shown that providing small group recitations in large lecture General Chemistry courses can improve student success with higher grades, a higher pass rate, and increased retention (Perera et al., 2019; Mahalingam et al., 2008). Further, it has been shown that recitations can be especially valuable for underrepresented students (Stanich et al., 2018). Virginia Tech has for many years utilized peer-led, small group recitations in CHEM 1036, our second semester General Chemistry course. These recitations provide active learning, an opportunity for one-on-one instructor-student interactions, and a community learning environment; our research over numerous semesters has shown that students performed better in CHEM 1036 after the introduction of the recitations. However, resources such as classroom space and funding for recitation instructor salaries are insufficient to include similar recitations for all CHEM 1035 students in the fall semester when our enrollment is 50% higher.

Efforts were therefore focused on using available resources to offer a few sections of a new course, CHEM 1034 General Chemistry Recitation. This one credit, pass/fail course, taken in conjunction with CHEM 1035, offers the advantages of the small group recitation. While any student may self-enroll in CHEM 1034 at the beginning of the semester, we found that the majority of students do not anticipate a need for the extra help provided by the recitation. The challenge, therefore, is identifying the students who will most need and benefit from participating in the recitation course. To do this, we used grades on the first test in CHEM 1035 to identify a target group of students who were invited to enroll in CHEM 1034 in the sixth week of the semester.

———————

METHOD

ENROLLMENT PROCEDURE

The first test in CHEM 1035 is given during the fifth week of the semester. Students whose Test 1 scores were in the range of thirty-seven to sixty-five received an email invitation from their professor to join a CHEM 1034 General Chemistry Recitation section. We found that students whose Test 1 score was above sixty-five would usually succeed in General Chemistry without the recitation course and that students whose scores were below thirty-seven on Test 1 usually immediately dropped the course. Those students who wanted to join the recitation course were added to a section by our Dean's office because students may not add a class themselves after the first week of the semester. The availability of four sections at different times in the late afternoon meant that most students could fit a section into their already established class schedules. In the Fall 2020 semester, the CHEM 1034 sections were taught in a synchronous, on-line format due to pandemic restrictions, while the Fall 2021 CHEM 1034 sections met in person. The enrollment of each recitation section was capped at a maximum of thirty students. We were able to accommodate all students who wanted to join a CHEM 1034 recitation.

RECITATION DESIGN

Each recitation section is taught by an undergraduate science major who has very successfully completed at least three semesters of chemistry (general and organic chemistry). The undergraduate recitation instructors are paid and are supervised by a faculty member.

The CHEM 1034 recitation class met once a week for seventy-five minutes. The focus of the weekly meeting was structured problem solving with students completing a worksheet in recitation under the guidance of the instructor; students were encouraged to engage each other and the instructor during this activity. A quiz with problems similar to those on that week's worksheet was given so students could gauge their understanding of the material. Worksheets and quizzes were written by the supervising faculty member.

ANALYSES

Data for both the Fall 2020 and Fall 2021 semesters were analyzed. We compared subsequent CHEM 1035 test grades and overall course grades of CHEM 1034 students (with recitation) to students in the same Test 1 grade range who elected not to enroll in CHEM 1034 (without recitation). In the Fall 2020 semester, a total of three tests were given, while four tests were given in the Fall 2021 semester.

A five-question evaluation survey was administered to the CHEM 1034 Recitation students in the Fall 2020 semester to solicit their opinions on the value of the recitation class.

RESULTS AND DISCUSSION

GRADE STATISTICS

We followed the progress of CHEM 1034 recitation students during the Fall 2020 and Fall 2021 semesters, comparing their subsequent test grades in CHEM 1035 General Chemistry with those students who scored in the same grade range (>37 to < 65) on Test 1 and opted not to enroll in CHEM 1034 recitation. As a result of pandemic policies during the Fall 2020 semester, only three tests were given in General Chemistry CHEM 1035. During the Fall 2021 semester, a total of four tests were given.

Table 3.1, Table 3.2, and Figure 3.1 compare the test averages of students scoring greater than 65 on Test 1 in CHEM 1035 with the students whose Test 1 score was in the range of > 37 to < 65 and who either opted to enroll in CHEM 1034 recitation or opted out of the recitation.

Table 3.1: Comparison of fall semester CHEM 1035 test averages (2020)

Cohort (2020)	Test 1 Average	Test 2 Average	Test 3 Average	Test 4 Average
Students scoring ≥ 65 on Test 1 (N = 1670)	83.2	77.5	75.5	NA
Students with recitation (N = 93)	60.0	70.3	70.4	NA
Students without recitation (N = 704)	53.2	57.1	62.1	NA

Table 3.2: Comparison of fall semester CHEM 1035 test averages (2021)

Cohort (2021)	Test 1 Average	Test 2 Average	Test 3 Average	Test 4 Average
Students scoring ≥ 65 on Test 1 (N = 1803)	84.4	78.5	80.0	66.9
Students with recitation (N = 85)	50.4	59.5	69.8	62.4
Students without recitation (N = 410)	54.1	59.6	65.5	58.9

Figure 3.1: CHEM 1035 test averages for students whose Test 1 scores were 65 or higher, students whose Test 1 scores were in the range of >37 to <65 and who opted to enroll in the CHEM 1034 recitation class after Test 1 (With Recitation), and students whose Test 1 scores were in the range of >37 to <65 and who did not enroll in CHEM 1034 after Test 1 (Without Recitation)

In the Fall 2020 semester:

- 12% of the CHEM 1035 students who were eligible to enroll in CHEM 1034 recitation on the basis of their CHEM 1035 Test 1 score opted to do so.

- The Test 2 and Test 3 averages were significantly higher for students with recitation compared to students without recitation.

Test average for students with Test 1 score ≥ 65 – Test average for students **with** recitation	Test average for students with Test 1 score ≥ 65 – Test average for students **without** recitation
Test 1: 23 points	Test 1: 30 points
Test 3: 5 points	Test 3: 13.4 points

In the Fall 2021 semester:

- 17% of the CHEM 1035 students who were eligible to enroll in CHEM 1034 recitation on the basis of their Test 1 score opted to do so.

- The Test 1 average was lower for students who enrolled in recitation (after Test 1) than for the students without recitation, but by Tests 3 and 4, the recitation cohort earned higher test averages than the students without recitation.

Test average for students with Test 1 score ≥ 65 – Test average for students **with** recitation	Test average for students with Test 1 score ≥ 65 – Test average for students **without** recitation
Test 1: 34 points	Test 1: 30 points
Test 4: 4.5 points	Test 4: 8 points

The subsequent CHEM 1035 test grades for the students whose CHEM 1035 Test 1 scores were >37 and <65 were examined to determine the amount of progress or lack of progress for students enrolled in the CHEM 1034 recitation versus the students who opted not to enroll in recitation. Test 2, Test 3 (and for Fall 2021, Test 4) grades were compared to Test 1 grades for each student. The results are shown in Table 3.3, Table 3.4, and Figure 3.2.

Table 3.3: Comparison of fall semester CHEM 1035 test grades for students with recitation versus without recitation (2020)

Cohort (2020)	Test 2 – Test 1 Grades	Test 3 – Test 1 Grades	Test 4 – Test 1 Grades
Students with recitation (N = 93)	Average change: +10.3 pts Test 2 ≥ Test 1: 79.6%	Average change: +10.3 pts Test 3 ≥ Test 1: 79.5%	NA
Students without recitation (N = 704)	Average change: +3.9 pts Test 2 ≥ Test 1: 66.2%	Average change: -2.8 pts Test 3 ≥ Test 1: 55.1%	NA

Table 3.4: Comparison of fall semester CHEM 1035 test grades for students with recitation versus without recitation (2020)

Cohort (2021)	Test 2 – Test 1 Grades	Test 3 – Test 1 Grades	Test 4 – Test 1 Grades
Students with recitation (N = 85)	Average change: +9.3 pts Test 2 ≥ Test 1: 82.2%	Average change: +18.0 pts Test 3 ≥ Test 1: 89.8%	Average change: +10.1 pts Test 4 ≥ Test 1: 76.5%
Students without recitation (N = 410)	Average change: +5.8 pts Test 2 ≥ Test 1: 68.5%	Average change: +11.4 pts Test 3 ≥ Test 1: 81.6%	Average change: +4.6 pts Test 4 ≥ Test 1: 62.2%

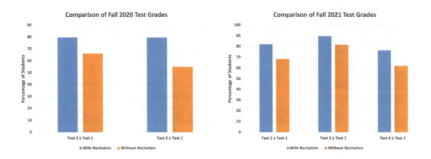

Figure 3.2: The percentage of students whose subsequent test scores were greater than or equal to their CHEM 1035 Test 1 scores

A larger percentage of students in recitation had an increase in subsequent test scores than did the students who were not in recitation in both the Fall 2020 and Fall 2021 semesters. It is important to note that, as shown in Table 3.1 and Figure 3.1, CHEM 1035 test averages for students scoring $\geq$ 65 on Test 1 typically decrease as the semester progresses.

In the Fall 2020 semester, students in the CHEM 1034 recitation increased their CHEM 1035 Test 2 and Test 3 scores by an average of 10.3 points, while students who were not enrolled in the CHEM 1034 recitation had an average increase of 3.9 points on Test 2 and an overall average decrease of 2.8 points on Test 3.

In the Fall 2021 semester, students in the CHEM 1034 recitation increased their CHEM 1035 Test 2, Test 3, and Test 4 scores by an average of 9.3, 18.0, and 10.1 points respectively, while students who were not enrolled in the CHEM 1034 recitation had an average increase of 5.8, 11.4, and 4.6 points on Test 2, Test 3, and Test 4, respectively.

The overall letter grade and pass rate of the two cohorts of students, those in CHEM 1034 recitation and those who were not, were compared. Since C– is the minimum acceptable grade in CHEM 1035 for several majors at Virginia Tech, the percentage of students earning that minimum grade is also reported. The results are shown in Table 3.5 and Table 3.6.

Table 3.5: Comparison of fall semester CHEM 1035 final grades (2020)

Cohort (2020)	Average CHEM 1035 Grade	% Passed CHEM 1035 (≥ D-)	% ≥ C-
Students with recitation (N = 93)	1.91	77.4%	57%
Students without recitation (N = 704)	1.19	77.5%	21%

Table 3.6: Comparison of fall semester CHEM 1035 final grades (2020)

Cohort (2021)	Average CHEM 1035 Grade	% Passed CHEM 1035 (≥ D-)	% ≥ C-
Students with recitation (N = 85)	1.72	91.0%	66%
Students without recitation (N = 410)	1.43	82.2%	48%

A higher percentage of students who enrolled in the CHEM 1034 recitation completed CHEM 1035 with a passing grade (≥D–) and with the minimum grade of C– required by several majors at Virginia Tech, than students who were not enrolled in the recitation course. Recitation students also earned a higher average letter grade in CHEM 1035.

STUDENT PERCEPTIONS

A five-question evaluation survey was administered to CHEM 1034 recitation students to solicit their opinions on the value of the recitation course. Figure 3.3 displays the results of the surveys.

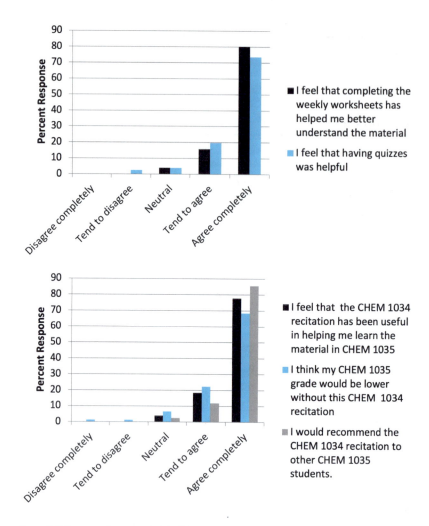

Figure 3.3: Summary by percent of the 76 responses to the CHEM 1034 Survey questions

Written student comments on the evaluation survey indicated a high level of satisfaction with the CHEM 1034 recitation course, as seen in a representative sample of the comments:

Sample of student comments

- "I believe this recitation course helped me get a passing grade in CHEM 1035 and I would absolutely recommend it to all students in that course. A lifesaver!"

- "The course was very helpful to me because it reinforced material that I had learned in the lectures and gave me a chance for extra practice. I would not change anything, I really enjoyed how I had the ability to ask any questions and that my instructor would help with it."

- "At first, I was not sure if I wanted to enroll in CHEM 1034 and commit to the "extra" work every week. However, I knew it would be good for me and I decided to enroll in the class. I am beyond grateful of this decision. The worksheets and recitations were easy to complete and took little time in comparison the tremendous amount of help I received. I enjoyed having scheduled study times and the additional exposure to more chemistry material every week."

- "The course was extremely helpful and definitely contributed to my improvement in Chem 1035. I will recommend this to anyone I meet who is going to take chemistry or is struggling in chemistry. I would not change anything about the course, I believe it is designed very well."

- "Helpful: – Walking through many practice problems step by step – Accepting questions at any step in a process – Verbally explaining each step. I went from a 60%ish on test 1 stressing if I would even pass chem but got above an 80% on test 2 and 90% on test 3. I'd say anyone can improve their chem 1035 grade with chem 1034."

CONCLUSION

CHEM 1035 students who enrolled in the CHEM 1034 recitation course after earning a low grade on the first test in CHEM 1035 had overall higher averages on subsequent tests and a higher overall average letter grade in CHEM 1035 at the end of the semester than CHEM 1035 students with a low first test grade who opted not to enroll in CHEM 1034. Furthermore, a higher percentage of the students in the recitation course earned the minimum grade of C– that is required by several majors at Virginia Tech, with the result that fewer of the students in recitation had to repeat General Chemistry to progress in their majors.

Survey responses and comments indicated a high level of satisfaction with the CHEM 1034 recitation course. It is troublesome that less than 20% of the eligible students chose to enroll in CHEM 1034. In the email invitation to join CHEM 1034, we told the students that we had noticed that they had struggled on Test 1, and they could take advantage of an opportunity to improve their grades on future tests. We presented data from the previous year that illustrated the increase in test grades and letter grades that many CHEM 1034 students experienced. However, this information did not seem to motivate the majority of the eligible students. Several sections of CHEM 1034 were offered in a variety in time slots; while some students may not have been able to fit the course into their class schedules, many students who opted not to enroll indicated that they thought the recitation course would be extra work and that they could improve their performance in CHEM 1035 on their own.

REFERENCES

Mahalingam, M., Schaefer, F., & Morlino, E. (2008). Promoting student learning through group problem solving in general chemistry recitations. *Journal of Chemical Education, 85*(11), 1577-1581.

Perera, V. L., Wei, T., & Mlsna, D. A. (2019). Impact of peer-focused recitation to enhance student success in general chemistry. *Journal of Chemical Education, 96*(8), 1600-1608.

Stanich, C. A., Pelch, M. A., Theobald, E. J., & Freeman, S. (2018). A new approach to supplementary instruction narrows achievement and affect gaps for underrepresented minorities, first-generation students, and women. *Chemistry Education Research and Practice, 19*(3), 846-866.

This program is supported in part by a grant to Virginia Tech from the Howard Hughes Medical Institute through the Inclusive Excellence Grant.

CHAPTER 4.

ASSESSING CHANGES IN STUDENT ENGAGEMENT USING A MIXED-METHODS APPROACH

JAMIE K. LAU; JEANNE MEKOLICHICK; AMANDA C. RAIMER; AND SARAH A. KENNEDY

ABSTRACT

Our Howard Hughes Medical Institute (HHMI) Inclusive Excellence program, REALising Inclusive Science Excellence (REALISE), focuses strategically and simultaneously on faculty development, curricular reform, student support, and institutional change. Assessment of our program is instrumental in determining the program's success in changing student perceptions of themselves as scientists. We employed a mixed-methods approach to explore how REALISE changes the ways in which Biology, Chemistry, and Physics students perceive their ability uncertainty, science identity, sense of belonging, and institutional commitment. First, we quantitatively explored the relationship between ability uncertainty and science identity as they predict sense of belonging. Then, we had an opportunity to qualitatively explore how the COVID-19 pandemic may have exacerbated any differences among underrepresented groups for these constructs. We found that female students were more uncertain in their abilities during the pandemic. Further, the relationship between sense of belonging and institutional commitment was weaker during the pandemic. Interestingly, few students felt like their campus belonging was affected by the pandemic. Our chapter provides successes and lessons learned while we developed and used our Student Engagement Survey.

INTRODUCTION

Unfortunately, between 48 and 56% of students who enter post-secondary education as a science, technology, engineering, and mathematics (STEM) major do not earn a STEM degree (Chen, 2013). This attrition is not a reflection of our incoming students, but instead a reflection of what typically greets them – a cold social and professional networking environment (Lane, 2016) or unconscious bias toward students' perceived ability to succeed (Steele, 2010). Like other environments, science contexts and cultures are constructed via implicit and explicit signals of exclusion or welcoming. We shape, and are shaped by, the contexts within which we live. Science identities, and decisions tied to the development of those identities, are cultivated (or not) within these environments via various social psychological constructs (Mead, 1934). As we strive to support all of our STEM students from their first day on campus to graduation, STEM education research continues to explore the social environment in which students engage with their peers and faculty, including the impact that these experiences have on sense of self, belonging, and success (Merolla et al., 2012; Merolla and Serpe, 2013; Stets et al., 2017). Indeed, cultivating a science identity and sense of belonging can be critically important drivers of success for all STEM students (Lane, 2016; Stets et al., 2017; Nunn, 2021). Because faculty, staff, and students create and maintain the social environment, there is opportunity for us to change ourselves, our culture and structures to ensure an inclusive, welcoming science context, engaging classrooms, and supportive faculty mentors.

Radford University is a mid-sized public university that primarily serves undergraduate students. At the time of writing our HHMI-IE proposal, 31.7% of our 2014-2015 graduating class entered as a STEM major but did not graduate with a STEM degree (personal communication, Dr. Timothy Millard, Radford University Institutional Research, 6/21/2023), which does contribute to the low percentage (40%) of STEM degrees earned nationwide (Toven-Lindsey et al., 2015). In an attempt to reverse this trend, our HHMI-IE-funded REALising Inclusive Science Excellence (REALISE) program (Wojdak et al., 2020) focused strategically and simultaneously on faculty professional development on inclusive pedagogies (Kennedy et al., 2022), curricular reform (Lau et al., 2019; see also chapter "Community, Curriculum, and CURES" in this book), student support (see the chapter Amplifying Student Voice and Vignette later in this book), and institutional change (see the chapter "Institutionally Advancing Inclusive Excellence" in this book) in the Biology, Chemistry, and Physics departments. These aspects of the program aimed to reduce student ability uncertainty in their courses, while developing a science identity and sense of belonging to Radford University, especially among our underrepresented students. Assessment is instrumental in determining the program's success in changing student perceptions of themselves as scientists. Thus, we designed a Student Engagement Survey (SES) to explore:

1. How the REALISE program changed the ways in which students perceive their ability uncertainty in their major, science identity, sense of belonging to the university, and institutional commitment; and

2. How students' social, academic, and college sense of belonging was affected by the COVID-19 pandemic.

To our knowledge, our model is one of the first to analyze the complex interactions among student demographics, uncertainty, science identity, belonging, and commitment within the context of assessing a program designed to change the students' perceptions of their abilities and science identity. We expected to see a reduction in ability uncertainty leading to an increase in science identity, sense of belonging, and institutional commitment. This chapter provides the successes and lessons learned while we developed and used the SES to measure the effect of REALISE on students in these three departments.

THE STUDENT ENGAGEMENT SURVEY (SES)

Quantitative and qualitative data are instrumental when thinking about student engagement. Quantitative data give us "what is happening," while qualitative data tell us "why" is the pattern is happening. Drawing on extant literature, we designed our Student Engagement Survey (SES) to include five constructs (i.e., concepts): ability uncertainty, science identity salience, science identity prominence, sense of belonging, and institutional commitment. We selected published instruments because the reliability in measuring each construct was validated. This kind of validation allowed us to immediately begin collecting data and assessing our program's outcomes (versus designing a new instrument that can take years to validate). Further, aligning our SES to the published instruments gave us the ability to compare our results to other studies at different institutions.

We used Lewis and Hodges's (2015) **ability uncertainty (AU)** scale to measure the degree of uncertainty about one's status as an able, competent member in a group. This instrument consists of twelve statements (e.g., I'm not sure that I'm cut out for my major) rated on a six-point Likert scale; higher scores equal more uncertainty about one's ability.

The **science identity salience (SIS)** scale came from Merolla et al. (2012), Piatt et al. (2019), and Stryker (2003), which measures the ranking of an identity on an internal hierarchy (e.g., how likely are you to tell a coworker, friend, or family member about the desire to be a scientist); more meaningful identities are more salient and more likely to be invoked in more situations. The SIS consists of four statements that are ranked on an 11-point Likert scale (from zero to ten); higher scores indicate higher SIS.

We used Stets' et al. (2017) **science identity prominence (SIP)** scale to measure the degree to which a science identity is a fundamental, central, or peripheral part of the self (e.g., I have come to think of myself as a "scientist"). The SIP consists of four statements that are rated on a five-point Likert scale of agreement; higher scores indicate higher SIP.

We modeled the **sense of belonging (SoB)** scale after Hausmann et al. (1990), which measures the degree to which individuals feel valued, needed, and significant in a group, system or environment (e.g., I feel a sense of belonging to Radford University). The SoB consists of four statements that are rated on a five-point Likert scale, from which we separated the **institutional commitment** measure (i.e., I am confident I made the right decision to attend Radford University). Higher scores indicate higher SoB.

During the height of the COVID-19 pandemic, we asked an additional question: how has the pandemic impacted your college experience? We framed the qualitative analysis of the participants' responses in three ways: social, academic, and college belonging (Nunn, 2021). We classified statements that disrupted social belonging if the participant mentioned student-led organizations or friendships with shared experiences/interests. Disruptions of academic belonging were classified when participants mentioned study groups, engaging with professors, or academic challenges. We classified statements about the physical environment, university-wide programming, or climate as a disruption of campus belonging.

DATA COLLECTION

Students' perceptions on each construct were collected via an Institutional Review Board (IRB)-approved instrument created from the scales listed above and delivered electronically using Qualtrics software. A link to the SES survey was distributed to faculty in our College via email during the mid-point of each semester. Faculty then shared the survey link with students in their courses. This method provided a cross-sectional collection of the data, from students taking science courses in that semester. As such, we did not necessarily sample the same students each semester. For example, the percentage of females and white participants represented in our sample was the highest in Fall 2022 (10% higher than in Fall 2020, Table 4.1a and Table 4.1b). First-generation participants fluctuated between 5 and 7% each sampling year. The percentage of REALISE department participants was similar to non-REALISE participants in Fall 2022, unlike previous sampling years. We used a pathway analysis to determine the relationships among the identity constructs over time using data from three semesters: Fall 2018 (pre-REALISE), Fall 2020 (height of COVID-19), and Fall 2022 ("post"-REALISE). These years are being used as a demonstration of the analytical capability of this survey; we also collected the data in the Spring semesters as well.

METHODS

The faculty distributed our SES with an added bonus (determined by the faculty) to ensure a better response rate (preferably at least two hundred participants; Baldwin, 1989). As a result, some students may have taken the survey multiple times to earn the bonus points for multiple classes. Therefore, we removed repeated submissions, keeping the participant's first submission.

Participants may not always respond to each statement for a number of reasons (Byrn, 2016). Rather than exclude their voice based on a few missing responses (< 5% missing per participant), we decided to impute the data to ensure an inclusive dataset. This statistical method uses the relationships among the variables (including demographic data, Table 4.1a-Table 4.1d) to predict the likeliest response for the participant's missing data (Byrn, 2016). We recognize that our demographic data excludes some information about students (e.g., Pell-eligibility, transfer status, number of credits earned, among others); however, these demographics were available to us at the time of sampling. We imputed the dataset five times with twenty iterations per imputation, and the final "likeliest" response was averaged across the five imputations; the *mice* package in RStudio was used to complete the imputation (van Buuren and Groothuis-Oudshoorn, 2011).

Table 4.1a: Participant gender demographics

Gender	Fall 2018	Fall 2020	Fall 2022
Sample Size (count)	269	291	79
Female (%)	63.4	60.8	70.9
Male (%, Reference Group)	36.6	38.1	29.1

Table 4.1b: Participant ethnicity demographics

Ethnicity	Fall 2018	Fall 2020	Fall 2022
Sample Size (count)	269	291	79
American Indian or Alaska Native (%)	0.4	0.3	0.0
Asian (%)	2.6	3.1	0.0
Black or African American (%)	10.4	14.1	10.1
Hispanic (%)	7.1	6.9	7.6
Native Hawaiian or Other Pacific Islander (%)	0.4	0.0	2.5
Nonresident Alien (%)	1.1	0.3	0.0
Race and Ethnicity Unknown (%)	2.2	4.1	2.5
Two or more races (%)	4.8	7.2	3.8
White (%, Reference Group)	68.4	62.9	73.4

Table 4.1c: Participant first generation college student demographics

First Generation College Student	Fall 2018	Fall 2020	Fall 2022
Sample Size (count)	269	291	79
Yes (%)	34.6	27.5	32.9
No (%, Reference Group)	62.8	71.5	67.1

Table 4.1d: Participant REALIZE department member demographics

REALIZE Department	Fall 2018	Fall 2020	Fall 2022
Sample Size (count)	269	291	79
Yes (%)	65.1	53.6	50.6
No (%, Reference Group)	32.3	45.4	49.4

THE PATHWAY ANALYSIS

A pathway analysis allows us to look at both direct and indirect relationships among multiple variables. Our analyses were guided by the literature, which shares that female students are often more uncertain in their abilities than male students (Lewis & Hodges, 2015). Further, female or Black, Indigenous, and people of color (BIPOC) students identify less as scientists because of underrepresentation in the field (Piatte et al., 2019) or satisfying social relationships in STEM (Merolla & Serpe, 2013). Therefore, these demographic variables are predicting AU and both science identity measures (Figure 4.1). However, the literature is not definitive about how ability uncertainty, science identity, and sense of belonging relate.

One critical aspect of the REALISE program was faculty professional development in inclusive pedagogy (Lau et al., 2019; Kennedy et al., 2022). To the degree faculty are able to create a more welcoming environment by reducing ability uncertainty, students' science identity may increase. Further, the development of a science identity should increase a student's sense of belonging and subsequent institutional commitment (Figure 4.1). Thus, we tested this pathway using the *lavaan* package in RStudio (Rosseel, 2012); however, the AMOS software is a known, user-friendly program for pathway analyses. See Byrne (2016) as a guide. We analyzed the pathway for Fall 2018, Fall 2020, and Fall 2022 separately to understand any changes in the relationships over time.

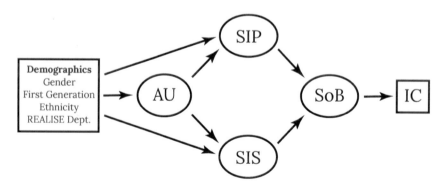

Figure 4.1: Simplified pathway analysis using our Student Engagement Survey

RESULTS

We can conclude that the direction of our proposed pathways is supported by the data by looking at the model fit statistics: comparative fit index (CFI), Tucker-Lewis index (TLI), and root mean square error of approximation (RMSEA). A pathway analysis is deemed a "good fit" when the CFI and TLI are greater than 0.95, and the REMSEA is less than 0.06. See Hu and Bentler (1999) for more information.

The data supported our proposed pathway analysis for Fall 2018 and Fall 2020 (Table 4.2). In other words, we can reliably interpret the relationships. However, the pathway for Fall 2022 is less reliable (Table 4.2). Unfortunately, the sample size was too small; we are interpreting the relationships with caution.

Table 4.2: Summary of model fit statistics per collection year

Metric	Fall 2018	Fall 2020	Fall 2022
Sample Size	269	291	79
CFI	0.966	0.968	0.907
TLI	0.960	0.961	0.891
RMSEA	0.046	0.047	0.080

WHAT INFORMATION CAN BE GATHERED FROM THE PATHWAY ANALYSIS?

Generally, our analysis confirms that an increase in students' ability uncertainty in their major reduces their science identity, and subsequently affects students' sense of belonging and commitment to the University – at least in Fall 2018 and 2020. Only during the height of the pandemic did female students have a 9% higher ability uncertainty (Table 4.3; 0.563/6-point scale) than male students. In Fall 2018 (pre-REALISE), students in Biology, Chemistry, and Physics (REALISE departments) had a 26.4% higher SIS (Table 4.3; 2.906/11-point scale). Interestingly, this relationship decreased at the height of the pandemic and decreased further in Fall 2022. Also, REALISE department students had a 16.5% higher SIP (Table 4.3; 0.823/5-point scale) than those in other departments. This relationship declined during the pandemic but seemed to recover in Fall 2022. This trend may have occurred because students were able to "do science" again in their coursework. Pre-REALISE, ethnically underrepresented students had an 8.5% lower SIS; however, this difference was not significant in Fall 2020 or Fall 2022. Only science identity prominence increased a student's sense of belonging by 4.9% (Table 4.3; 0.287/6-point scale) in Fall 2018 and 5.2% in Fall 2020. Interestingly, the relationship between sense of belonging and institutional commitment weakened during the pandemic (from 0.885 to 0.743) and began to strengthen again in Fall 2022 (from 0.743 to 0.755; Table 4.3).

Table 4.3a: Summary of pathway analysis; ability uncertainty
($* P < 0.05$; $** P < 0.01$; $*** P < 0.001$)

Ability Uncertainty	Fall 2018	Fall 2020	Fall 2022
Gender	0.220	0.563**	-0.506
First Generation	0.287	0.262	0.235
REALISE Department	0.074	-0.043	0.150
Ethnically Underrepresented	0.184	0.106	-0.320

Table 4.3b: Summary of pathway analysis; science identity salience
($* P < 0.05$; $** P < 0.01$; $*** P < 0.001$)

Science Identity Salience	Fall 2018	Fall 2020	Fall 2022
Gender	0.448	0.083	0.172
First Generation	-0.112	-0.640	0.392
REALISE Department	2.906***	2.288***	2.151**
Ethnically Underrepresented	-0.933*	-0.116	0.783
Ability Uncertainty	-0.352*	-0.395**	-0.508

Table 4.3c: Summary of pathway analysis; science identity prominence
(* P < 0.05; ** P < 0.01; *** P < 0.001)

Science Identity Prominence	Fall 2018	Fall 2020	Fall 2022
Gender	0.118	0.074	0.439
First Generation	0.038	-0.048	0.140
REALISE Department	0.823***	0.673***	0.889***
Ethnically Underrepresented	-0.213	-0.127	0.266
Ability Uncertainty	-0.160***	-0.178***	-0.075

Table 4.3d: Summary of pathway analysis; sense of belonging
(* P < 0.05; ** P < 0.01; *** P < 0.001)

Sense of Belonging	Fall 2018	Fall 2020	Fall 2022
Science Identity Salience	-0.032	-0.010	-0.015
Science Identity Prominence	0.287**	0.309***	-0.190

Table 4.3e: Summary of pathway analysis; institutional commitment
(* P < 0.05; ** P < 0.01; *** P < 0.001)

Institutional Commitment	Fall 2018	Fall 2020	Fall 2022
Sense of Belonging	0.885***	0.743***	0.755***

WHY DO SOME OF THESE TRENDS EXIST?

The changes in science identity were likely the result of students being removed from the campus environment and disconnected from their peers and academics. We were able to discover the "why" by framing our content analysis using Nunn's (2021) concepts of social, academic, and college belonging. Comments in bold relate to **social belonging**; comments in purple (underlined) relate to academic belonging; comments in orange (non-italicized in quotes) relate to campus belonging.

For example (quotes from STEM majors),

> I already have a **really hard time making friends and with the pandemic it has made it even harder**. Sometimes I have no motivation to do anything since I do not have anyone to talk to and that has *affected my school work* (Female, White, Freshman, Pell-eligible).

> I feel that the pandemic has *negatively impacted student engagement in a few of my courses*. I do feel that there is a *disconnect between classmates and professors*, but I appreciate how Radford has tried to hold **socially-distanced events** and has encouraged us to do our best to be a part of the Radford community. Sometimes, *I do feel isolated from peers and in-person learning would be more ideal for my learning experience*, but I think that online courses make me feel safer than I would if I were solely on campus (Female, Biracial, Senior).

> I feel **extremely limited socially**. I would have loved to have a normal first year, and I am anxious for my future at Radford University as I do not see the pandemic disappearing in the near future (Male, White, Freshman, Pell-eligible).

> I'm new here. I think the professors and admin have done a good job. *I'm glad to have class in person and to be able to study with my classmates while distanced*. I don't prefer online learning (Male, Black, Junior).

> I am the type of student that learns more by being in class rather than online. *My grades have been impacted (not in a good way)* and I haven't learned much this semester (Female, Hispanic, Pell-eligible, First-gen, Junior).

Interestingly, most of the participants mentioned a disruption in academic belonging, followed by social belonging. Only four participants felt like all three were disrupted (Figure 4.2). We were not surprised by the pandemic's ability to disrupt academic and social belonging – two of the primary reasons why students attend post-secondary education. Although post-secondary education is a time for academics, we were surprised that participants rarely mentioned campus belonging. The "college experience" stereotypically includes school pride (e.g., "go Highlanders", sporting events, intramural sports) and social engagements across campus (e.g., student clubs and organizations). Radford either did well at maintaining campus belonging during the pandemic or campus belonging is not cultivated in Radford students. We were not able to tease out this aspect of campus belonging from our dataset.

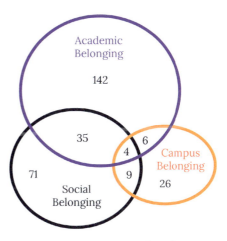

Figure 4.2: Venn diagram quantifying the types of college belonging that our participants mentioned in their responses

CONCLUSION

Our Student Engagement Survey aimed to capture the complexity of the student experience in our courses by blending ability, identity, and belonging into one theoretical framework (Lane, 2016). We were able to see who was affected most by the pandemic (female students) and how the relationships between ability uncertainty, science identity, and sense of belonging declined and cautiously recovered after the pandemic. Unfortunately, we cannot comment specifically on the success of REALISE using the survey because the pandemic introduced many factors unrelated to the aims of REALISE. However, given data on the positive impacts of implementing inclusive pedagogies, we hope that faculty who engaged in our program were able to reduce the pandemic's potential to exacerbate inequity in our STEM courses. Excitingly, the sense of belonging scale is being used as an assessment tool for

Radford's Quality Enhancement Program (QEP) that arose out of the REALISE program successes.

LESSONS LEARNED

In evaluating our approach to the data collection process and the content of the student engagement survey instrument, we recognize a few lessons learned. First, we should have measured the same students before and after implementation of an initiative to ensure a longitudinal collection of the data that better measures change in the participants over time. Second, in a future effort we would deploy the survey across the college or university to ensure that students taking the survey are not limited to the faculty who provide incentives for survey completion. Our deployment approach limited our ability to assess change at college, department, and class levels. As part of this approach, designing sense of belonging questions as they relate to the major in addition to the university (e.g., I feel a sense of belonging to my [insert major].) would have allowed for a more granular assessment of belonging. Ultimately, the pandemic created an enormous and disruptive challenge for our faculty and students in the middle of our grant program. We approached this unexpected event as an opportunity and modified our original SES to help gain information about how the pandemic was affecting our students. This pivot allowed us to learn new and different perspectives from our students.

Acknowledgements

We thank the Howard Hughes Medical Institute's Inclusive Excellence in STEM Education Grant (#52008708) for funding our REALISE program. Thank you to the faculty who distributed our SES and to the students who completed the SES. Thank you also to our anonymous reviewer for their comments that allowed us to clarify our work.

REFERENCES

Baldwin, B. (1989). A primer in the use and interpretation of structural equation models. *Measurement and Evaluation in Counseling and Development, 22*(2), 100-112.

Byrn, B. M. (2016). *Structural equation modeling with Amos* (3rd ed.). Routledge.

Chen, X. (2013). *STEM attrition: College students' paths into and out of STEM fields*. Statistical Analysis Report. NCES 2014-001. National Center for Education Statistics. https://files.eric.ed.gov/fulltext/ED544470.pdf

Hausmann, L. R., Ye, F., Schofield, J. W., & Woods, R. L. (2009). Sense of belonging and persistence in White and African American first-year students. *Research in Higher Education, 50*(7), 649-669.

Huston S., Herman R., Liss S., & Taylor B. (2023). Community, curriculum, and CUREs: Transformations in the physics department at Radford University. In J. Briganti, J. Sible, & A. M. Brown (Eds.), *Fostering communities of transformation in STEM higher education: A multi-institutional collection of DEI initiatives* (pp 103-118). Virginia Tech Publishing.

Kennedy, S. A., Balija, A. M., Bibeau, C., Fuhrer, T. J., Huston, L. A., Jackson, M. S., Lane, K. T., Lau, J. K., Liss, S., Monceaux, C. J., Stefaniak, K. R., & Phelps-Durr, T. (2022). Faculty Professional Development on Inclusive Pedagogy Yields Chemistry Curriculum Transformation, Equity Awareness, and Community. *Journal of Chemical Education, 99*(1), 291-300.

Lane, T. B. (2016). Beyond academic and social integration: Understanding the impact of a STEM enrichment program on the retention and degree attainment of underrepresented students. *CBE Life Sciences Education, 15*(3), ar39.

Lau, J. K., Paterniti, M., & Stefaniak, K. S. (2019). Crossing floors: Developing an interdisciplinary CURE between an environmental toxicology course and an analytical chemistry course. *Journal of Chemical Education, 96*(11), 2432-2440.

Lewis, K. L., & Hodges, S. D. (2015). Expanding the concept of belonging in academic domains: Development and validation of the ability uncertainty scale. *Learning and Individual Differences, 37*, 197-202.

Mead, G.H. (1934). *Mind, self, and society from the standpoint of a social behaviorist.* University of Chicago Press.

Mekolichick J. (2023). Institutionally advancing inclusive excellence: Leading from the middle in times of transition. In J. Briganit, J. Sible, & A. M. Brown (Eds.), *Fostering communities of transformation in STEM higher education: A multi-institutional collection of DEI initiatives* (pp 219-230). Virginia Tech Publishing.

Merolla, D.M., & Serpe, R.T. (2013). STEM enrichment programs and graduate school matriculation: The role of science identity salience. *Social Psychology Education, 16*(4), 575-597.

Merolla, D.M., Serpe, R.T., Stryker, S., & Schultz, P.W. (2012). Structural precursors to identity processes: The role of proximate social structures. *Social Psychology Quarterly, 75*(2), 149-172.

Nunn, L. M. (2021). *College belonging: How first-year and first-generation students navigate campus life.* New Jersey: Rutgers University Press.

Piatt, E., Merolla, D., Pringle, E., & Serpe, R.T. (2019). The role of science identity salience in graduate school enrollment for first-generation, low-income, underrepresented students. *The Journal of Negro Education, 88*(3), 269-280.

Robinson, K. A., Perez, T., Carmel, J. H., & Linnenbrink-Garcia, L. (2019). Science identity development trajectories in a gateway college chemistry course: Predictors and relationship to achievement and STEM pursuit. *Contemporary Educational Psychology, 56*, 180-192.

Rosseel, Y. (2012). lavaan: An R package for structural equation modeling. *Journal of Statistical Software, 48*(2), 1-36.

Steele, C. M. (2010). *Whistling Vivaldi: How stereotypes affect us and what we can do*. W. W. Norton & Company.

Stets, J.E., Brenner, P.S., Burke, P.J., & Serpe, R.T. (2017). The science identity and entering a science occupation. *Social Science Research, 64*, 1-14.

Stryker, Sheldon. [1980] (2003). *Symbolic interactionism: A social structural version*. Blackburn.

Toven-Lindsey, B., Levis-Fitzgerald, M., Barber, P.H., & Hasson, T. (2015). Increasing persistence in undergraduate science majors: A model for institutional support of underrepresented students. *CBE Life Science Education, 14*(2): ar12.

van Buuren, S., & Groothuis-Oudshoorn, K. (2011). mice: Multivariate imputation by chained equations in R. *Journal of Statistical Software, 45*(3), 1-67.

Wojdak, J., Phelps-Durr, T., Gouch, L., Atuobi, T., Deboy, C., Moss, P., Sible, J., & Mouchrek, N. (2020). Learning together: Four institutions' collective approach to building sustained inclusive excellence programs in STEM. *Transforming institutions accelerating systemic change in higher education*. Pressbooks. http://openbooks.library.umass.edu/ascnti2020/chapter/wojdak-etal/

This program is supported in part by a grant to Virginia Tech from the Howard Hughes Medical Institute through the Inclusive Excellence Grant.

CHAPTER 5.

DEVELOPMENT AND ASSESSMENT OF A FOUR-WEEK SUMMER RESEARCH EXPERIENCE FOR UNDERGRADUATES

DEBORAH J. GOOD AND ERICA ECHOLS-MILLER

INTRODUCTION

Implementing and sustaining undergraduate research programs is a top priority for higher education, as such programs contribute to short- and long-term student success (Eagan et al., 2013). Many experts suggest that significant improvement in undergraduate education and graduate school enrollment would come from requiring undergraduate research and inquiry-based learning (Eagan et al., 2013). Furthermore, multiple lines of evidence suggest that summer research programs can specifically increase the number of underrepresented minority students entering graduate programs and health professions ((Bruthers & Matyas, 2020; Prince et al., 2023; Quintana, 2021). Most summer research programs are nine to twelve weeks in duration. However, during the summer 2022 period, we piloted a unique four-week summer experience for six undergraduate students from varying science, engineering, technology, and mathematics (STEM) disciplines. The program was called the Inclusive Excellence (IE) Fellows. Student selection was based on those individuals who had traditionally been underrepresented in summer experiences and who had barriers to participating in longer summer research programs. Pre- and post-survey data showed that IE Fellows had self-perceived gains in communication skills, creativity, autonomy, dealing with obstacles, inquiry, and disciplinary knowledge, although the sample size was too low to have statistical significance. IE mentors agreed with self-perceived Fellow results for communication, inquiry, and disciplinary knowledge gains. Analysis of open responses from a focus group of participants found that use of a weekly elevator pitch practice allowed students to improve their communication skills, and confidence in their research project, which complement the gains in skills measured in the survey. In summary, even though the four-week period was short, the IE Fellows program can serve as a model for a short, inclusive summer research immersion for students and faculty who have barriers to longer-term summer programming.

After reading this chapter, individuals will:

1. Have the tools to implement a 4-week summer research immersion.
2. Understand the advantages and challenges of short-duration summer research programs.

BACKGROUND

The opportunity to do undergraduate research in the sciences literally changed my life (D.J.G). Throughout my academic career, an ongoing goal has been to provide more undergraduate research opportunities for students—especially those who may not get the chance to engage in authentic research experiences, due to various barriers, including those whose socioeconomic status prevented the extra time for undergraduate research, minoritized students who may not feel like undergraduate research programs were for them, or students from first-generation, or non-traditional backgrounds who did not know how an undergraduate research experience may benefit them. The Inclusive Excellence (IE) Fellows program was started with these thoughts in mind, following an unexpected windfall of money from the Office of Research at Virginia Tech, which "had to be spent" that year. By combining this windfall with money from the Howard Hughes Medical Institute (HHMI) Inclusive Excellence program sub-fund for the Department of Human Nutrition, Foods, and Exercise, we were able to come up with a program where five students could be given a $2,000 scholarship for the spring semester, and a $1,000 fellowship for the four-week program. Additional money was set aside to induct each student who presented into the Sigma Xi Scientific Research Honor Society (www.sigmaxi.org) ($60 each).

Finding money to fund these types of programs without a federal or foundation grant can be easier than one might imagine, but one must have a budget ready, and a need articulated when money becomes available. As an example, in past years, I (D.J.G.) found funds for another 10-week summer program that I run (now as an National Institutes of Health-funded program) through university institutes and department "slush" funds. Institutes and departments generally have programmatic money and your pitch for getting this money for your undergraduate research program should include information on your target student group, your program milestones, how you will track or report the outcomes of the program, and whether there are plans to seek federal or foundation funding in the future. The institute or department then knows how your program's outcomes will benefit their reporting requirements. A good approach is to be specific in providing the dollar amounts needed for your program (including other sources of support), and to not be afraid to approach any group on campus that might be willing to support your program from their coffers. Generally, those groups, institutes and departments have money that they must distribute and need to find good programs to give to. Programs increasing experiential learning opportunities for undergraduates have always been readily supported when the funds available signify a win-win for both sides.

The short-term nature of the IE Fellows was developed for three reasons. First, the money we had available would not be sufficient to match that of the longer-term programs (usually $5000 or more per student over a ten-week program). Second, I (D.J.G.) wondered if a short-term summer research program would be appealing to students who had not had an opportunity to participate in a program prior to this, or who had time limitations, or other barriers to undergraduate research program participation. There are no other short-term undergraduate research programs (less than five weeks) that have been described and published based on searches of the ERIC or PubMed databases. However, McLaughlin et al. (2018) showed significant impacts for a short-term research experience embedded in a seventeen-day study abroad program. Third, my colleague and I who run a ten-week program were feeling the pain of "mentor-burnout", and wondered if a shorter program may also benefit mentors. Mentor burnout syndrome is a documented psychological condition in university professors caused by a combination of emotional exhaustion, depersonalization, and low professional accomplishment (Redondo-Florez et al., 2020). Interestingly, a search of the ERIC and PubMed databases for information on mentor burnout in undergraduate research programs did not reveal any published studies on the issue. However, anecdotally, we have heard from mentors that the issue is real. As the academic calendar tends to leave summer session open for many faculty to focus on their own research or personal pursuits, shorter summer research programs may help provide that time for faculty mentors.

Diversity, in the context of the IE Fellows program included differences and variables among participants including, but not limited to race, ethnicity, gender, sexual orientation, socioeconomic status, age, ability, religion, culture, and nationality. It also encompassed the unique characteristics, experiences, and perspectives that IE Fellows brought to the community of scholars. This program strived to recognize that every person is different and value those differences as important contributions to the program (and more broadly the campus community and society). As shown in Table 5.1, Embracing Diversity was one of the first topics discussed in the program. To us, "Embracing Diversity" meant fostering an inclusive and equitable environment where everyone is respected, valued, and given the opportunity to succeed and contribute.

When mentoring across cultural boundaries, it is important to recognize commonalities and differences in the mentor/mentee relationship. Cross-cultural mentoring strategies in higher education are crucial for promoting diversity, equity, and inclusion (DEI) on college and university campuses. Mentoring relationships between individuals from different cultural backgrounds can provide opportunities for sharing knowledge and increasing mutual understanding. Some effective cross-cultural mentoring strategies include building trust through open and honest communication, providing training and resources to develop the mentee's skills and competencies, and setting clear goals and expectations that support the mentee's personal, professional, or academic development (Chan, 2010). By implementing each of these strategies, higher education institutions can promote cross-cultural understanding and help students from diverse backgrounds succeed academically and professionally.

IMPLEMENTATION

Recruitment of interested students from Virginia Tech STEM disciplines occurred using email list-servs and by word of mouth. All undergraduates in Inclusive Excellence Departments at Virginia Tech (College of Science, College of Natural Resources and the Environment, and College of Agriculture and Life Sciences) were invited to apply. Funds to support this program were obtained from the Office of Research ($2,000 Spring scholarship per student, $10,000 total), the HHMI grant ($1,000 summer fellowship per student, $5,000 total), and induction into the Sigma Xi Research Honor Society ($60 per student, $300 total). While there was no formal requirement for the program in the spring, the source of the monies determined when they were distributed. In total, students directly received a $3000 stipend for their four-week program. There was a very short turnaround time between announcement of the program and application deadline (six days) due to the nature of the Spring scholarship funds which had to be distributed during the Spring academic semester. Students had to have identified a research mentor in advance, with the mentor writing a short statement agreeing to host the student for the program if they were selected. To identify students who might benefit from the IE Fellows program, emails were sent specifically to faculty currently participating in the HHMI-funded Inclusive Excellence program, as well as to the list-serv for the Office of Undergraduate Research (OUR). Students participating in a department DEI book club were personally invited to apply by D.J.G. Eight students representing all three STEM colleges applied, and six students were selected by a team of two faculty reviewers, including D.J.G., using a ranking rubric (available on request). D.J.G. excused herself from ranking her own student during this process.

The program ran from July 5, 2022-July 29, 2022, culminating in the (OUR) Summer Research Symposium, on July 28th, 2022. Students were expected to spend eighty hours doing research, participating in weekly group meetings (two hours per week), completing surveys (pre/post), participating in one focus group, and presenting a poster at the full day OUR Summer Research Symposium. The scheduled topics for the weekly group meetings are shown in Table 5.1. Each IE Fellow and mentor pair was asked to complete a survey prior to the start of the program and within one week of the end of the program. In addition, IE Fellows participated in a one-hour focus group the one week after completing the program. The surveys and focus group questions were approved as "exempt" by the Institutional Review Board.

Table 5.1: Weekly meeting topics. References and videos used to facilitate some of these topics are included.

Week Number	Topics
Week 1	<ul><li>Introductions</li><li>Group icebreaker (https://www.thoughtco.com/ice-breaker-the-name-game-31381)</li><li>DEI in science-what does it mean, how do we obtain it? (Levin, 2021; Urbina-Blanco et al., 2020)</li><li>Elevator pitch discussion and practice (LetPub)</li></ul>
Week 2	<ul><li>Science Communication and Networking (LinkedIn, Twitter, ResearchGate discussion)</li><li>Elevator pitch practice</li><li>Abstracts and conference registration due</li></ul>
Week 3	<ul><li>Elevator pitch practice</li><li>Present draft posters to group (discussion based on (Hess, 2013; Toven-Lindsey et al., 2015; Urbina-Blanco et al., 2020))</li><li>Final poster due for printing</li></ul>
Week 4	<ul><li>Practice poster presentation</li><li>Sigma Xi Scientific Research Honor Society Membership Application</li><li>Office of Undergraduate Research Summer Research Symposium, full day</li></ul>

RESULTS AND DISCUSSION

Participant demographics: There was a short application period of ten days, based on the timing of receiving information about the extra money and confirmation that we could go ahead with the proposed plan. Students were also required to identify a mentor and provide a short letter from this individual stating that if they were accepted to the program, the mentor would provide a summer research experience for them in July. Due to the short application time frame, only eight individuals applied to the program. The program applicants represented diverse majors, including Nutrition, Biochemistry, Biology, Sustainable Biomaterials, Animal Science, and Wildlife Sciences. A total of six students were selected as IE Fellows, based on a scoring rubric that took into account diversity status (race, ethnicity, gender, sexuality, disability status, veteran status, economic status, first-generation status), previous research experience, any self-reported barriers to undergraduate research (including not being previously selected for research positions, work during the summer etc.), and reasons for wanting to join the program. Grade point average or academic honors were not collected in the application. Table 5.2a outlines the self-reported participant demographics for the IE Fellows. As shown, the IE Fellows were diverse in several categories, with each student having at least one demographic that designated them as part of an underrepresented group in undergraduate research programs (Cooper et al., 2019; Gin et al., 2022; Sellami et al., 2021; Singer & Weiler, 2009; Toven-Lindsey et al., 2015). The relatively high number of underrepresented individuals in this group were likely due to the "call for applications" document which required the student to have at least one demographic that put them in a underrepresented minority (URM) category.

Table 5.2a: Participant demographics

Participant 1	Participant 2	Participant 3	Participant 4	Participant 5	Participant 6
female	female	transgender male	female	female	female
junior	freshman	senior	freshman	junior	senior
white	mixed race	white	mixed race	white	white
not first gen	Not first gen	first gen	parents were first gen	parents were first gen	first gen
doing research	not doing research	not doing research	not doing research	doing research	doing research
not disabled	not disabled	disabled	not disabled	not disabled	not disabled
veteran status-parent	not a veteran	not a veteran	not a veteran	veteran status-parent	veteran status-parent
economically disadvantaged	not economically disadvantaged	not economically disadvantaged	not economically disadvantaged	parents were economically disadvantaged	economically disadvantaged

Table 5.2b: Mentor demographics

Mentor 1	Mentor 2	Mentor 3	Mentor 4	Mentor 5
mentor to participant 1	mentor to participant 2	mentor to participant 3	mentor to participant 4	mentor to participant 5 and 6
female	male	male	male	female
white	Asian	Hispanic	white	white
first gen	not first gen	first gen	not first gen	first gen
not a veteran	not a veteran	not a veteran	not a veteran	not a veteran
not economically disadvantaged	parents were economically disadvantaged	not economically disadvantaged	not economically disadvantaged	not economically disadvantaged

Mentor demographics were also surveyed. Mentors (N=5, with one mentor having two IE Fellows) were white (3), Hispanic (1), and Asian (1); male (3) and female (2), with three mentors reporting that they were first generation college students. None of the mentors indicated that they or their parents were veterans. One mentor self-reported that they came from an economically disadvantaged background. Sexuality data was not collected from mentors.

Survey Findings: During the application process, participants described their barriers to conducting research as financial limitations, time, and access, including, "I do not receive any financial aid for courses during the summer"; "I use the summer to get extra hours at my job to cover expenses during the semester"; "Transphobia [in my discipline leads to] social and mental barriers to my authentic participation in research"; and finally, "There have been times I have nearly had to quit doing my research in order to get a job, but have continued doing my research because I am so passionate about it. I hope that through this [program] I will not have to worry about my financial situation so much and will be able to focus on my schooling and research more."

Once accepted to the program, IE Fellows were given a pre-survey. The pre-survey re-asked the demographic questions (Table 5.2a) and asked them to comment on their expectations for the program. For the expectations questions (open comments), participants indicated that they wanted to "become a more proficient communicator"; "learn from others in the program", "meet new peers"; "understand how research operates at different levels"; "to gauge my love of work in the lab"; "to meet with other diverse scholars and discuss research"; "to work in a collaborative environment"; "to improve my research methods"; and "to learn how to create a diverse and inclusive environment in research."

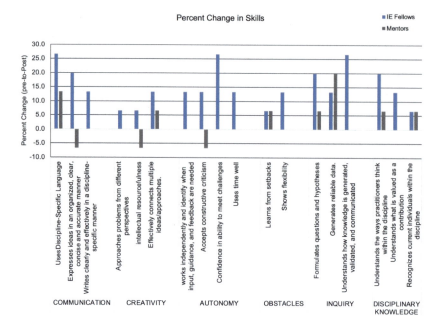

Figure 5.1: Skill acquisition by IE Fellows. IE Fellows and their mentors were asked to complete a skill acquisition survey prior to starting their summer experience, and within one week of completing it. Only three IE Fellows and mentors (of five) completed both surveys. Data shown is derived from a Likert scale analysis (1 – extremely effective and 5 = not effective at all) converted to percentage. IE Fellows (blue bars) ranked their own skills, while mentors ranked their Fellow's score in each category, with percent change shown (gray bars).

Each IE Fellow and their mentor then took a validated undergraduate research assessment survey (Singer & Weiler, 2009). The undergraduate research assessment instrument listed six groupings of questions targeted to determine skill levels in communication, creativity, autonomy, obstacles, inquiry, and disciplinary knowledge. The assessment survey portion was given both before the program started and after completion of the four-week program, to both the IE Fellows and their mentors, who were asked to rank their IE Fellows in each category. The results of the pre- and post-survey data show that the IE Fellows scored themselves as having gains in all categories. Generally, mentors' scores for their IE Fellows were lower in each of the categories (Figure 5.1). Note that lack of bars in the graph indicates there was no change pre-to-post survey for that category. Questions from communication skills, inquiry skills, and disciplinary knowledge showed stronger mentor-rated gains than in creativity or obstacles, although the sample size was too low to show significance, and some students or mentors did not complete both pre- and post- surveys.

The pre-survey also asked IE Fellows to comment on their concerns in the week prior to the program start. Two of the Fellows indicated that they had no concerns, one was concerned about whether they could get everything done in just four weeks, another was concerned about what it meant to be part of a program, and the last was concerned about the fact that their own project was done in Brazil, and whether they would be able to connect with their peers.

Focus Group Findings: Dr. Erica Echols at the National Institute for STEM Evaluation and Research (NISER) conducted a focus group to understand outcomes of student participation in the IE Fellows program. Five out of six students participated in the focus group. Preliminary review of open responses from the focus group found that use of a weekly elevator pitch practice allowed students to improve their communication skills, and confidence in their research project. IE Fellows also shared that even though the four-week period was short, they had increased their knowledge of laboratory practices, improved their time management, and learned how to balance their schedule by planning research and meeting other deadlines.

Table 5.3a: Thematic analysis of NISER focus group with IE Fellows: Have you gained any disciplinary knowledge you would not have gained in the classroom alone?

Theme	Theme Definition
Networking and resources	This theme is characterized by the discussion of different websites, such as ResearchGate and Google Scholar, and the importance of setting up profiles on these sites to connect with others and access research materials.
The value of research lab experience	This theme highlights the differences between classroom and laboratory experiences and how the latter can provide a deeper understanding of the scientific process. IE fellows found the research lab experience to be more informative and practical than classroom learning and appreciated the opportunity to apply classroom knowledge in a practical setting.
Transfer of knowledge	This theme underscores the importance of applying classroom learning to real-life situations, as it allows students to see the practical value of what they have learned. The research lab experience was particularly valuable in this regard, as it helped IE fellows understand why seemingly abstract concepts and information are important.
Fieldwork	This theme describes the importance of hands-on experience in understanding certain aspects of science, particularly fieldwork. IE fellows felt that some things cannot be learned in the classroom and that fieldwork experience is necessary to fully appreciate and understand certain aspects of scientific research.

Table 5.3b: Thematic analysis of NISER focus group with IE Fellows: Did meeting with the other IE fellows improve your program experience? If so, how?

Theme	Theme Definition
Collaborative Learning	The theme of collaborative learning emphasizes the importance of learning with and from others in a group setting. Students emphasized how meeting with other IE fellows from different disciplines allowed for broader learning and research opportunities. This collaborative learning environment provided opportunities for practicing and honing research skills, which were enriched through the exchange of ideas and feedback.
Sharing and Communication	The theme of sharing and communication highlights the importance of open and effective communication in learning and research. IE fellows emphasized the importance of being able to share research experiences and struggles with others who have similar experiences, as it fosters a sense of community and support.
Interdisciplinary Learning	The theme of interdisciplinary learning highlights the value of learning from individuals with diverse backgrounds and interests. IE fellows shared how the interdisciplinary nature of the other students in the program allowed for learning across different scientific fields, which offered fresh perspectives and unique insights.
Personal Growth	The theme of personal growth emphasizes how collaborative learning and interdisciplinary learning environments can lead to personal growth and development. IE fellows noted how being part of the program allowed for growth in terms of research skills, confidence, and social skills.

Table 5.3c: Thematic analysis of NISER focus group with IE Fellows: How has your mentor impacted your summer experience?

Theme	Theme Definition
Conference Preparation	IE fellows received training and preparation for conference presentations. Mentors held presentation rehearsals, and one student was able to attend a university-funded research trip.
Learning about the financial aspect of academia	IE fellows gained understanding of how research funding is awarded in academia, as well as insight into the research process from a financial perspective.
Mentorship and learning about lab resources	IE fellows received mentorship and guidance from experienced researchers and were exposed to different resources available in a lab setting.

Thematic analysis (Strauss & Corbin, 1990) was also conducted on the focus group responses for three of the focus group questions (Tables 5.3a, 5.3b, and 5.3c). One theme that IE Fellows highlighted was the importance of practical experience and networking in scientific research, as well as the value of applying classroom learning to real-life situations. This theme also underscores the limitations of classroom learning and the importance of hands-on experience in fully understanding certain aspects of science. Second, IE Fellows highlighted the value of collaborative learning and interdisciplinary learning in scientific research. Their responses also emphasized the importance of open and effective communication and the role of such communication in fostering a sense of community and support. These responses reinforce how learning environments such as the IE Fellows program can lead to personal growth and development. Third, research mentors impacted the IE Fellows' research experience through exposure to available researcher resources and providing opportunities to present their research at a conference.

Mentor Perspectives: In the pre-survey, mentors were asked to comment on their expectations for the program, their concerns about the program, and the length of the program in the pre- and post-surveys. In the pre-survey, the mentors' expectations for the program included providing training on presenting research, learning more about the research process, and having their IE Fellow interact with others from diverse backgrounds. Following the program, mentors commented that the program was organized, and that the presentations and meetings seemed beneficial to the IE Fellows. One mentor specifically commented that the group meetings helped get their student "out of her shell." Mentors also reported that the students developed confidence and learned about research possibilities within their field. For concerns about the program prior to the start, only one mentor indicated that they were concerned that they would not be able to connect with their student in such a short period of time. Following the program, the mentor wrote *"I mentored a transgender woman student that has suffered discrimination and has been unable to participate in research projects in the past due to their identity and economic limitations. The IE Fellowship program was the only opportunity available to allow this student access to this learning opportunity in my research group."* The post-survey comments from mentors indicated concern about the short length of time for the program, and mentors felt that students needed to have a research project in place, or they would not be able to have a poster presentation in that short period of time. A small amount of progress was made on research projects, but all students had poster presentations with new data from the summer. All mentors indicated that they would be willing to host their IE Fellow in the fall, and the three students who were still enrolled as undergraduates continued with these mentors in the Fall. This program did not provide direct mentor or laboratory resources.

LESSONS LEARNED

Several lessons learned during this summer research program can be used by others who may want to try the shorter duration summer program.

1. Surveys and focus groups helped to define what worked and what could be improved in the program, with the survey results generally supporting the result found using the student focus groups. Interestingly, the focus group responses indicated students wanted more time in group meetings during the summer (a single one-hour meeting was held per week), and some planned social group outings. Increasing group interactions, especially outside of the training, is one way to build relationships with the program director and the other fellows, which could increase

group cohesion (Harden et al., 2014).

2. Shirts bearing the Inclusive Excellence logo (Figure 5.2) were provided to each participant at the end of the semester. The use of a group name and associated logo is an effective strategy to increase group cohesion and distinctiveness (Estabrooks et al., 2012). It is recommended that programs, no matter their duration, have a unique logo that can be put on shirts or other program material. The short timeframe for the IE Fellows program meant that students did not receive their shirt during the program. Thus, it was not possible to determine if the logo and group name indeed fostered group cohesion.

A.  B.

Figure 5.2: Inclusive Excellence Fellows and shirt logo A. Four of the six members gathered in Fall 2022 wearing their IE shirts with program director, Deborah Good (center). Faces are partially blocked for the IE Fellows to protect their identity B. A picture of the IE Fellows logo

3. This study provided evidence that shorter, more focused summer research programs may benefit diverse students by providing a short-term experience that still allows the student to do other things (classes, work, travel) during the summer. For faculty directing a shorter-term program, there are lower cost in both faculty time, and money to run program. Student fellowship can be less expensive for shorter duration programs., For example, a typical 10-week program pays a fellowship of $5250. A comparable four-week program would cost $2100 per student. Our program paid slightly more ($2000 scholarship and $1000

summer fellowship) based on the source of our funds.

4. There were concerns from mentors that four weeks was not enough time to have data for a poster presentation, and indeed, three of the six students who joined the IE Fellows were already doing undergraduate research in the previous semester. Thus only two of the five students (40%) were new undergraduate researchers. The shorter programs may require that students spend the semester prior to the program in a research group (perhaps with a scholarship or fellowship in the Spring semester, paying for their research time). Alternatively, they could be selected in January, and then begin conversations with the mentor to define the project prior to the start in the summer. One of the students who was not yet doing undergraduate research leveraged the program to gain admittance to their research group. This was one of the best outcomes from the program as the student then continued with the research group in the following Fall.

5. We suggest the following timeline for faculty who are considering implementing a shorter summer program:

    1. Fall applications. This would allow students who did not want to directly approach a faculty mentor to be "matched" with faculty already willing to participate in the program.

    2. A one-credit spring semester course (one to three hour per week) in undergraduate research. This course could include interaction with the mentor and other members of the research team during their lab meetings, as well preparation for the short-term project that could be completed during the summer program. Students might read journal articles and discuss these with their mentors, or even participate in some hands-on training in the Spring semester in advance of the summer.

    3. Short summer program culminating with presentation (poster or oral presentation). For the IE Fellows, the programming was scheduled to culminate on the week where the Office of Undergraduate Research hosted their Summer Research Symposium. All IE Fellows presented their posters at the symposium.

4. One follow-up group meeting in the following semester. This could be a group activity or a lunch and would continue to foster group cohesiveness.

In summary, the short-term summer research program for six students was a success. All students who did not graduate were still working with their research groups the following semester. Ideas for improving the program, such as starting with a spring course, and including at least one group outing are provided for faculty interested in implementing a short-term undergraduate research experience in their units.

REFLECTIVE QUESTIONS

Based on our experience, a few reflective questions were designed to help as you consider building a short-term summer research program with cross-cultural mentoring and inclusive excellence in mind.

Developing a 4-week summer research experience:

- How would students in my department benefit from a 4-week summer research experience? Survey students in your courses to see if there is interest in summer research and what barriers exist that may hinder their participation. Maybe they are taking summer courses or need to work during the summer. A one-month summer commitment to research may be more conducive to many students' schedules and allow exposure and growth.

- What do I, as a faculty mentor, value more in regard to student outcomes: quantity of research data obtained, or the experience provided to the student?

- Where can I leverage money to support a short-term summer research program?

- Who are your target student populations for this program?

Cross-cultural mentoring:

- List the top three things you value most in your professional/academic environment.
- Ask your mentee to also list the top three things they value most in their professional/academic environment.
- Discuss your values with your mentee and share why these values are in your top three.
- Discuss any common or differing values and steps that can be taken to improve in these areas.
- Resources
  - Chan, A. (2010). Inspire, empower, connect: reaching across cultural differences to make a real difference. Lanham, MA, Rowan & Littlefield Education. (Chan, 2010)
  - Mathews, P. (2003). Academic mentoring: enhancing the use of scarce resources. Educational Management & Administration. 31(3), 313-333. (Mathews, 2003)

Inclusive excellence:

- What steps have I taken as a research mentor to ensure that diverse student populations feel valued and included in the research community?
- How can I help (or continue to help) these students build networks and connections that will support their future academic and professional success?
- Resources
  - Karten, T. J. (2015). Inclusion strategies that work!: Research-based methods for the classroom. Corwin Press. (Karten, 2015)

REFERENCES

Bruthers, C. B., & Matyas, M. L. (2020). Undergraduates from underrepresented groups gain research skills and career aspirations through summer research fellowship. *Advances in Physiology Education, 44*(4), 525-539. https://doi.org/10.1152/advan.00014.2020

Chan, A. (2010). *Inspire, empower, connect: reaching across cultural differences to make a real difference*. R&L Education.

Cooper, K. M., Gin, L. E., Akeeh, B., Clark, C. E., Hunter, J. S., Roderick, T. B., Elliott, D. B., Gutierrez, L. A., Mello, R. M., Pfeiffer, L. D., Scott, R. A., Arellano, D., Ramirez, D., Valdez, E. M., Vargas, C., Velarde, K., Zheng, Y., & Brownell, S. E. (2019). Factors that predict life sciences student persistence in undergraduate research experiences. *PLoS One, 14*(8), e0220186. https://doi.org/10.1371/journal.pone.0220186

Eagan Jr, M. K., Hurtado, S., Chang, M. J., Garcia, G. A., Herrera, F. A., & Garibay, J. C. (2013). Making a difference in science education: The impact of undergraduate research programs. *American Educational Research Journal, 50*(4), 683-713. https://doi.org/10.3102/0002831213482038

Estabrooks, P. A., Harden, S. M., & Burke, S. M. (2012). Group dynamics in physical activity promotion: What works? *Social and Personality Psychology Compass, 6*(1), 18-40. https://doi.org/10.1111/j.1751-9004.2011.00409.x

Gin, L. E., Pais, D., Cooper, K. M., & Brownell, S. E. (2022). Students with disabilities in life science undergraduate research experiences: Challenges and opportunities. *CBE—Life Sciences Education, 21*(2), ar32. https://doi.org/10.1187/cbe.21-07-0196

Harden, S. M., Estabrooks, P. A., Mama, S. K., & Lee, R. E. (2014). Longitudinal analysis of minority women's perceptions of cohesion: The role of cooperation, communication, and competition. *International Journal of Behavioral Nutrition and Physical Activity, 11*(1), 57. https://doi.org/10.1186/1479-5868-11-57

Hess, G. R. (2013). *Giving an effective poster presentation* [Video]. Youtube. https://www.youtube.com/watch?v=vMSaFUrk-FA

Karten, T. J. (2015). *Inclusion strategies that work!: Research-based methods for the classroom.* Corwin Press.

Going Up! Elevator Pitches for Scientists. LetPub. https://www.letpub.com/ author_education_Going_Up#:~:text=An%20elevator%20pitch%20or %20speech,ride%20an%20elevator%20between%20floors.

Levin, A. G. (2021). *How to begin building a culture of diversity, equity, and inclusion in your research group.* AAAS. https://www.science.org/ content/article/how-begin-building-culture-diversity-equity-and-inclusion-your-research-group

Mathews, P. (2003). Academic mentoring: Enhancing the use of scarce resources. *Educational Management & Administration., 31*(3), 313-333.

McLaughlin, J., Patel, M., Johnson, D. K., & de la Rosa, C., L. (2018). The impact of a short-term study abroad program that offers a course-based undergraduate research experience and conservation activities. *Frontiers: The Interdisciplinary Journal of Study Abroad, 30*(3), 100-118.

Prince, L. Y., Williams, T. B., Allen, A. R., McGehee, R. E., Jr., & Thomas, B. R. (2023). Outcomes of the UAMS summer undergraduate research program to increase diversity in research and health professions. *Advances in Physiology Education, 47*(1), 20-25. https://doi.org/ 10.1152/advan.00201.2022

Quintana, D. S. (2021). Replication studies for undergraduate theses to improve science and education. *Nature Human Behaviour, 5*(9), 1117-1118. https://doi.org/10.1038/s41562-021-01192-8

Redondo-Florez, L., Tornero-Aguilera, J. F., Ramos-Campo, D. J., & Clemente-Suarez, V. J. (2020). Gender differences in stress- and burnout-related factors of university professors. *BioMed Research International, 2020,* 6687358. https://doi.org/10.1155/2020/6687358

Sellami, N., Toven-Lindsey, B., Levis-Fitzgerald, M., Barber, P. H., & Hasson, T. (2021). A unique and scalable model for increasing research engagement, STEM persistence, and entry into doctoral programs. *CBE—Life Sciences Education, 20*(1), ar11. https://doi.org/10.1187/cbe.20-09-0224

Singer, J., & Weiler, D. (2009). A longitudinal student outcomes evaluation of the Buffalo State College summer undergraduate research program. *CUR Focus, 29*(3), 20-25.

Toven-Lindsey, B., Levis-Fitzgerald, M., Barber, P. H., & Hasson, T. (2015). Increasing persistence in undergraduate science majors: A model for institutional support of underrepresented students. *CBE—Life Sciences Education, 14*(2), ar12. https://doi.org/10.1187/cbe.14-05-0082

Urbina-Blanco, C. A., Jilani, S. Z., & Speight, I. R. (2020, August 17). *Science is everybody's party: 6 ways to support diversity and inclusion in STEM.* World Economic Forum. Retrieved April 7, 2023, from https://www.weforum.org/agenda/2020/08/science-stem-support-inclusion-diversity-equality/

This program is supported in part by a grant to Virginia Tech from the Howard Hughes Medical Institute through the Inclusive Excellence Grant.

CHAPTER 6.

CREATING IMPACTFUL MOMENTS

Using Peer Role Models to Build Community and Sense of Belonging in STEM
AMANDA C. RAIMER; KRISTINA STEFANIAK; KAITLYN EDWARDS;
JYNNA HARRELL; TRERESE ROBERTS; AND SANDRA LISS

Kaitlyn Edwards, Jynna Harrell, and Trerese Roberts acted as peer role
model authors who wrote their reflections on the program, seen in
quotes, and assisted with reading the chapter for content.

ABSTRACT

A unique aspect of Radford University's Howard Hughes Medical
Institute Inclusive Excellence (HHMI IE) grant involved training peer
role models (PRMs) to engage early–career college students in biology,
chemistry, and physics majors beyond the classroom. The PRM program
has entered its sixth year and undergone many iterations along the way.
The successes and challenges encountered have resulted in a framework
that other institutions can use in creating their own peer role model
program. In this chapter, we describe the importance of student voice
in the program's success and the various ways we've built community
extending beyond the three original departments to encompass all of
science, technology, engineering, and mathematics (STEM). Additionally,
we provide a framework for the successful responsibilities of both the
faculty mentors and the students in organizational meetings,
implementation of events, and professional development. Many
successful outcomes have been accomplished through PRMs, and we
hope to see similar models in other institutions.

INTRODUCTION: REALISE, SENSE OF BELONGING, AND PEER MENTORING

Radford University's HHMI REALISE (REALising Inclusive Science Excellence) grant aimed to increase student success and sense of belonging among biology, chemistry, and physics majors. Radford University is a mid-sized public university in southwestern Virginia focused primarily on undergraduates with increasing enrollment of groups historically excluded from STEM. Over the five years leading up to the grant (2013-2017), Radford University saw a 21.5% increase in Pell-eligible students, 22.4% increase in first-generations students, and 53.9% increase in underrepresented minority students (URMs). Within the three participating departments during the grant period (2018-2022), 38.8% of the students were Pell-eligible students, 35.5% were first-generation students, and 34% were URMs. Additionally, similar to other institutions, Radford experiences lower first- to second-year retention rates in STEM majors. Many factors contribute to lower retention rates and this grant specifically focused on supporting students' sense of self, exploring the social constructs of ability uncertainty, self-efficacy, science identity, and sense of belonging.

Sense of belonging can be conceptualized as including three sub-categories: academic belonging, social belonging, and campus belonging (Nunn, 2021). Academic belonging is a student's feeling of belonging within their classes, study groups, and other academic clubs/organizations. Social belonging is a student's feeling of connectedness and inclusion with peers outside of academics. Campus belonging is the student's connectedness to their institutional community and environment as a whole. REALISE's assessment of STEM students' sense of belonging during the COVID-19 pandemic showed that students most often mentioned impacts to their academic and social belonging, showing the importance students placed on these aspects of their college experience (Lau et al., 2023). While multiple aspects of REALISE focus on increasing academic belonging through faculty and curriculum development, the REALISE Leadership Team designed a unique aspect of our grant that involved training peer role models (PRMs), or REALISE Students, to engage early-career college students in our College of Science and Technology (see the chapter "Community, Curriculum, and

CUREs" in this book; Kennedy et al., 2022). Peer mentoring has been shown to improve sense of belonging and academic success within STEM communities, and our PRM program was designed intentionally to augment both academic and social belonging (Clements et al., 2022; Zaniewski & Reinholz, 2016).

The leadership team recognized the importance of a peer-mentor component in order to keep student perspective as a clear voice in the evolution and direction of REALISE. Additionally, there are stark differences between our faculty and student demographics, so the PRMs brought a much-needed diverse set of backgrounds and experiences to the grant. Beyond the student perspective, the PRMs represented a spectrum of student voices as first-generation students, transfer students, student-athletes, commuters, and other identities as well as a variety of genders and races/ethnicities. Unlike most other PRMs or mentors who are associated with a specific course or major, our PRMs performed cross-departmental service that promoted community and sense of belonging for all of our STEM majors through programming primarily outside of the classroom. The development and implementation of each initiative was predominantly student-driven, and each semester's work was adapted to that specific group's goals for the program. The PRMs and their faculty mentors truly operated as a team to work through challenges and achieve their goals.

WHY DOES IT MATTER? GOALS OF THE REALISE PRMS

The PRM component was integral in creating an inclusive environment within the biology, chemistry, and physics majors. The purpose of our PRM program was twofold. First, they provided an avenue for the leadership team to hear about the student experience within the college. If you are trying to improve the student experience, then the students are the most important perspective. From these conversations, different programmatic REALISE initiatives were created. Their second role was larger; they were on the front line working to (1) build science identity, (2) increase science community, and (3) amplify student voice within the college, all done through the lens of diversity, equity, and inclusion (DEI).

The main goal of the REALISE grant was to increase STEM students' success through cultivating science identity and sense of belonging, both of which are critical drivers to student success and persistence in STEM (Estrada et al., 2011; Lane, 2016; Stets et al., 2017). One example of PRM-developed programming that focused on both of these drivers is the classroom visits, which we will discuss in detail later. From conversations between PRMs, mentors, and STEM students, we heard about some student barriers to success, like having a difficult time navigating the collegiate system, facing food insecurity, time management struggles with school plus work, and other similar themes. The classroom visits and events gave the PRMs time to engage with students; conversations were tailored around known barriers to our students. Another important outcome of the PRM program and their events was to make diversity in science more visible. Below is a statement from a STEM student turned PRM that highlights how the events hosted by the PRMs helped her build community in addition to increasing her science identity.

Student PRM Statement on Community and Science Identity

"When I graduated with my Bachelor's degree in 2018, I realized that I wanted to become a dentist. I decided to enroll into Radford University as a Biomedical Sciences major to complete my prerequisites that I needed to apply. My first semester was one of my hardest because my workload was almost entirely intense science classes. Creating relationships with my peers was the last thing on my mind because I was consistently comparing myself to others.

Partway through the semester I was walking down the hall our science building and noticed an eye- catching flier about an imposter syndrome seminar. I decided to go to the event because I wondered if maybe I was going through imposter syndrome myself. The event was hosted and managed by the REALISE Students, some of who I recognized from my classes. After this event I was able to meet the president of the Pre-Dental Club who was also a REALISE Student. She would invite me to other events that the program had which allowed me feel more comfortable around a variety of STEM majors.

I decided to go to more REALISE events because I enjoyed the topics that they covered. One of my favorite events was the diverse panel of scientists who came to talk to us about their experience of being a minority in the STEM field. I believe that events like that are what truly made me want to become even more active in the program. I went to some of their events to learn even more about the program and the student mentors in STEM just like me. One day I thought to myself that I would like to be a REALISE Student Mentor because I wanted to be a part of the mission of helping other students like myself who may have been having a hard time going through semester with such hard classes. I was persistent when it came to inquiring about and joining the program because I knew that I could offer even more support of people who look just like me."

-Trerese Roberts (Biomedical Sciences major)

The multi-faceted impacts of the PRM program are seen in experiences like Trerese's. The importance of the PRMs was highlighted during the COVID-19 pandemic when our students saw an abrupt decline in interactions with peers and faculty which resulted in a lost sense of community. Our PRMs faced the challenge of a greater need for sense of belonging and community while trying to completely change the format of their programming and keep it effective. The PRMs' student perspective, creativity, and resilience led to completely new or modified events that remained aligned with their goals. For example, the PRMs really leaned into using social media and our learning management system to share resources and connect with first-year students. These types of initiatives could still be used for PRMs that want impactful experiences that require less time and funding.

HOW DOES IT WORK? IMPLEMENTING A CROSS-DEPARTMENTAL, STUDENT-DRIVEN PRM PROGRAM

The PRM program is now in its sixth year at Radford University and during that time we've implemented and revised its structure. The following framework promotes a high level of voice and involvement from the PRMs which is key to reaching students across the college. It also acknowledges the importance of guidance and support from faculty mentors. The model we found to be most successful is a collaborative one based on three pillars: (1) weekly communication through meetings; (2) college wide initiatives to build community and sense of belonging; and (3) professional development focused on DEI and science identity for the PRMs. Figure 6.1 outlines the roles of the faculty mentors and the PRMs in each pillar.

Figure 6.1: The three main pillars of the PRM program and the roles of the faculty mentors and students.

Meetings

Faculty: Weekly mentor meetings to set goals for the week

Students: Weekly meetings with planning, advertising, and assessment sub-groups

Initiatives

Faculty: Purchasing and organizing off-campus visitors

Students: Developing event ideas, planning, advertising, running the event, and assessment

Training

Faculty: Identify areas of training and organize

Students: Participate and reflect

At the start of each semester, the PRMs met and decided on their goals for the upcoming semester. The mentors started the discussion by summarizing the overall programmatic goals of the REALISE program before turning it over to the PRMs. They were given time to consider their priorities individually, before coming back together to consolidate them into three to five themes. Below is a sample list of the main objectives of the PRMs for a semester.

- Creating Community among STEM students
- Broadening student participation in events
- Involving STEM faculty
- Increasing avenues for advertising events and resources
- Recruiting a larger, more diverse team of PRMs

WEEKLY MEETINGS

At the end of every semester, we as faculty mentors surveyed the PRMs and asked about their personal successes and challenges in addition to the program's successes and challenges. The theme that always showed up as a challenge was "communication". We all know college students are busy, especially those who self-selected to be part of our PRM program. The best way we've found to encourage communication is through a group chat containing both faculty mentors and PRMs as well as weekly meetings. The group chat was helpful in keeping the team updated between meetings on what needs to be done and where to find resources. There was nothing innovative about the chat other than having students acknowledge they have read the messages by "liking" them. This helped accountability and let students feel seen if they asked a question or shared progress on a project.

The weekly meetings were critical to the success of the PRM program. Although this was a student-run program, the faculty members played a pivotal role in managing the logistics of the initiatives and keeping the overall goal of creating community and belonging within the STEM majors at the forefront of all initiatives. To accomplish this there was a separate weekly meeting just for the faculty mentors. During the mentor meeting, we discussed recent events and whether we believed they met the overall goals set by the PRMs for the semester. We also created an agenda for the upcoming student meeting. This typically included time for reflections on current and past initiatives, some action items for upcoming ones, and a short training or professional development exercise. Along with setting the structure and foundation for the PRM meetings, this additional meeting among the faculty mentors gave us a weekly time to troubleshoot, brainstorm, and confirm a plan for the week for which we were all united.

The PRM meetings themselves opened with the PRMs reflecting on the effectiveness of previous events and initiatives, specifically discussing the assessment measures they designed. It was useful to start the meeting with these reflections because it reminded us of the goals of the program and set the tone for the planning and training that occurred later in the meeting. Afterwards, we guided them through a professional development activity (details below) and then transitioned to working time for the coming events and initiatives.

We have found that the best way to assign tasks and have them completed is to break into sub-teams. Each team took ownership of their specific tasks and assigned a point person to contact. The teams were composed of three to six students; making sure it was a mix of experienced and first-time PRMs was ideal. We found that having advertising, planning, and assessment sub-teams was an effective structure to divide the workload. We allowed time (fifteen to thirty minutes) during each weekly meeting during which the sub-teams collaborated to decide on their to-do list for the week and assign ownership of each task. Again, the ideas, implementation, and success of the events were in the students' hands. The mentors were there for support, purchasing, and logistics. Below is a long-time PRM who discusses the role of the meetings; you will hear her voice throughout this Implementation section.

Student PRM Statement on the Role of Meetings

"I joined REALISE in January 2022, after learning about REALISE from my microbiology professor who recommended that I apply for the peer role model position, and I am currently in my third semester in REALISE. Throughout my REALISE career, I have been a part of the advertising and planning teams. During weekly meetings, REALISE students plan academic-related and fun stress-buster events to host throughout the semester. Weekly meetings consist of planning events, emailing collaborators about the events, creating sign-up sheets, preparing supply lists, and discussing the overall goals of the events. Throughout the planning process of each event, we always reflect back on our semester goals. Is this event meeting the semester goals? After the event, during our weekly meeting, we reflect on the outcomes of the event. Did the event meet the semester goals? If so, how do we expect that outcome in other events? If not, how do we improve for the future? These are all important discussions that take place during meetings, the behind-the-scenes of REALISE. REALISE is much more than events and weekly meetings though, REALISE aims to encourage all STEM students to find their sense of belonging in science, doing so by hosting events and maintaining a presence in the college. I celebrate being a part of REALISE because it is a program that actively incorporates DEI into the STEM community at Radford University."

– Jynna Harrell (Biology major)

COLLEGE-WIDE INITIATIVES

As discussed earlier, the PRM program had three main goals: (1) build STEM identity, (2) increase STEM community, and (3) amplify student voice. The avenues to accomplish these varied from one-time large events to informative flyers posted around the science building. Throughout the program, we worked to strike a balance between creating impacts and not overburdening the PRMs. Each semester there were two to four one-time events and weekly Fresh Fruit Fridays. Fresh Fruit Friday began early in the program and was able to accomplish all three of the program's goals. We set up a table offering coffee, fruit, and granola bars staffed by PRMs and faculty who would actively engage students on their way to and from class in our science building every Friday morning. Each week there was also a question or prompt to help the PRMs engage with their peers. We used this event to help promote other PRM, college, and university programming that aligned with the REALISE goals.

The other events and initiatives were more targeted toward one of the main goals. To encourage growth of students' STEM identity, we held science identity exploration events, developed a diversity in STEM panelist/roundtable series, and visited first- and second-year STEM classes. To create community within the college, we organized STEM club fairs, stressbuster events, Fresh Fruit Fridays, and created support and resource fliers for the science building. Last, we collected student thoughts through surveys and individual event assessments, which we later compiled to disseminate to faculty within the college. An initiative that takes place every semester was classroom visits, where the PRM reached out and offered their services to instructors. Jynna will explain further.

Student PRM Statement on Classroom Visits

"One way REALISE achieves its goal of promoting a sense of belonging in STEM students is through classroom visits. REALISE students reach out to science professors who teach first and second-year classes, inviting them to host the REALISE students in their classrooms. Classroom visits are chances for REALISE students to have one-on-one conversations with new and transfer students in STEM, as well as an opportunity for students to be heard. REALISE students come prepared to discuss academic and personal topics about their own experiences being student scientists.

My most memorable classroom visits were for two back-to-back visits for biology major introductory seminar classes. The first classroom visit I completed that day was with another REALISE student. As we began the conversation with the classroom, I became aware that my fellow REALISE student and I had complementing college experiences. I was a local commuter student who transferred from community college, she was a student-athlete from across the country. I was in STEM to pursue healthcare; she was in STEM to pursue pre-veterinary studies. I had no previous knowledge of careers in STEM before entering college, she came from a background of scientists. Talking about our different experiences in front of a class of new STEM students was encouraging to me. It was clear to the students in the classroom that although we were different, anyone could be a scientist. Although our stories may be different from the students' own stories, there was an important place for them in STEM.

After finishing these two classroom visits, I became aware that the experience made me develop a strong sense of science identity. I came to the classroom visits to help other students build their sense of belonging, but I walked away gaining just as much from the experience as I hoped the students in the classroom received. I realized that it was up to REALISE to help students build their science identity through experiences like these. Through this experience, REALISE taught me how to make spaces for everyone in STEM."

– Jynna Harrell (Biology major)

TRAININGS

The ability to navigate conversations about DEI can be challenging. It would be unfair to task our PRMs with the goal of fostering an inclusive environment without training on what that means. This looked different every semester, but the main professional development for the PRMs included external speakers that led an Inclusive Spaces Workshop as well as a Mental Health First Aid training, an internal university-run certification, and short training modules within our meetings led by the faculty mentors. Some examples of training modules during the weekly meetings focused on role-playing scenarios to help overcome student barriers, creating an identity chart, and discussion on student demographics at Radford University and within the College of Science and Technology. Additionally, because the PRMs interacted directly with students in the college, it was important they knew the goals of the REALISE program and how to communicate this with their peers. To solidify this, every semester we spent a meeting giving the PRMs time to develop an elevator pitch describing the PRMs and REALISE program. Here a PRM described how valuable the trainings were, particularly regarding DEI:

Student PRM Statement on the Value of the Training

"Prior to being a REALISE Student, I was not as well-versed in DEI. Through REALISE, I was able to attend trainings that taught me how to recognize an issue and how to respond to that issue. In a meeting, we practiced scenarios of what we would do if we encountered certain situations. This not only opened my eyes to see that these scenarios can truly happen, but it helped me dive deeper into the resources here at Radford University and know the faculty that I can trust and reach out to if any of the scenarios happens."

– Kaitlyn Edwards (Biology major)

WHAT HAVE WE LEARNED? FACULTY AND STUDENT REFLECTIONS ON PRM SUCCESSES, CHALLENGES, AND EVOLUTION

Throughout the five years of the REALISE program and the changing group of PRMs we have learned several lessons.

INITIATIVE SUCCESS

As faculty mentors, we have found that allowing for a significant part of our weekly meetings to be *working meetings* contributed to the success of the event. The students who served as PRMs often had many other commitments (classwork, athletics, research, etc.) that take up a significant amount of their time and energy. Using our already-scheduled meeting times to complete critical tasks ensured that the tasks (1) got done and (2) included input from other PRMs and the faculty mentors.

Planning: Giving the students enough time to plan events was key to their success. Even small events had many moving pieces.

- *Who*: Who is planning the event? Who is advertising the event? Who is staffing the event? Who is assessing the effectiveness/impact of the event? Who do we need to collaborate with/contact to help with the event? Who is our audience for the event?

- *What*: What activities will we offer? What supplies/facilities/resources do we need?

- *When*: When (time/day) will the event be held? Do we have PRMs available to staff the event? Are there any other initiatives, events, or activities that may conflict with our times, dates, spaces and/or target audience?

- *Where*: Where will the event take place? How much space do we need? What location(s) do we need to reserve? Is it in-person, online, or hybrid?

- *Why*: What is the overall purpose of the event? What are the measurable goals? What tools will we use to assess if these goals have been met?

- *How*: How will we advertise the event? How will we get our target audience to attend and engage with the initiative? How will we infuse community-building and sense of belonging into our initiative?

Kaitlyn talks more about the challenges and lessons she learned about event planning and delivery:

> ### Student PRM Statement on the Challenges and Lessons Learned About Event Planning and Delivery
>
> "During my time in the REALISE student program, I learned that communication and teamwork are VERY important in everything you do. With all of the students' schedules conflicting, we had to separate the group into two meeting times. In order to keep tasks and events organized, we all had to communicate and work together. This was done in a variety of ways

including Excel sheets for sign-ups and tasks, GroupMe, and meeting notes. Keeping information neat and in the right spot is key in order for everyone to be informed.

On another note, I found that hosting events was a larger challenge than expected. We all have different schedules and our available times are difficult to line up. Also, an event may not always have the end result that was hoped for. Just because an event didn't meet our goals doesn't mean it holds no value. We use outcomes like this to reflect and see what can be improved for the next event. Finding the good in something that didn't meet the expectations can be challenging because when time is used to plan and advertise an event and no one comes, it can be discouraging. But, as a REALISE student, I didn't let it drag me down. I jumped onto the next event and used the prior outcomes to better the upcoming event."

– Kaitlyn Edwards (Biology major)

Advertising: Getting our target audience to the event began with advertisements. We have found that a consistent, extended advertising schedule that spans multiple forms of communication helps us maximize our impact. Typically, this meant having initial advertisements shared at least three weeks before the event. Several avenues we explored for reaching students include:

- Posted flyers in the STEM building
- Social media posts and stories
- Half- and quarter-page handouts to be distributed at other events
- Department and college-wide emails
- Verbal invitations on the day of the event
- Extra credit opportunities from faculty

In our experience, these last two were the most effective at contributing to turnout at our events. The day-of verbal invitations were personalized and timely, while the opportunity for extra credit provided an additional incentive for busy students to prioritize attendance.

Assessment: The events we planned required time, effort, and resources from many people. It was important that we knew that the work being put in was contributing to meeting the goals of the organization. As discussed above, our primary goal was to increase student success by (1) building science identity; (2) increasing science community; and (3) amplifying student voice.

One metric we used to determine the success of our events was to track attendance statistics. For some events this was simply counting how many students were engaging with the event. For most, we also asked how many years the students have been at Radford University, what their majors are, what REALISE PRM events they have attended in the past, and if they have any future events they'd like to see. One place where faculty mentorship of the PRMs was critical was in developing assessments that clearly aligned with the semester goals. Many of the PRMs had little to no experience with assessment, so guiding them through this alignment and referring to the goals regularly was critical.

Another challenge the PRMs had to work through over time was how the format of the assessments affected the response rate. Several of our PRMs initially liked the idea of holding informal conversations with event attendees to determine its success because it would promote interaction and community-building. However, as the faculty mentors we found that in practice most PRMs either weren't comfortable enough approaching that many "random" students or only talked to a few of the many attendees. Similarly, when we tried digital post-event surveys to save paper and resources our response rate decreased greatly. We've found that the most effective way for the PRMs to actively collect data and get a high response rate is through physical paper surveys and sign-in sheets.

REALISE STUDENT IMPACTS

One unforeseen but exciting outcome of the PRM program was the benefits the PRMs saw for themselves. Not only did they recognize improvement in transferable skills, but most reflected on how being a PRM increased their own sense of belonging and community in our college. Trerese, Jynna, and Kaitlyn all mentioned how they feel the PRM program impacted them:

Student PRM Statements on the Program's Impact

"What I have learned from being a REALISE Student is that I am a part of a wonderful program is truly committed to building a community of empowered faculty and student learners that is welcoming and inclusive. I have truly aligned myself with the vision that REALISE has. I love that we deeply aspire to engage students and cultivate a culture of excellence so all students believe they can achieve. What I have gained from being a REALISE Student is that I am a part of a program that helps change mindsets of students to believe that they can."

-Trerese Roberts (Biomedical Sciences major)

"REALISE has taught me a lot about myself, allowing me to use my goal-oriented mindset to increase the goals of REALISE in our college. As a commuter and a transfer student, being involved in college outside of academics can present challenges. Typically, I am only on campus on days I have classes and some semesters that can only be three days a week. REALISE works with my schedule, gives me opportunities to be involved, teaches me how to use my voice, teaches me how to lead, and teaches me the importance of DEI. I am learning just as much in REALISE as I am in the classroom, and being a part of the REALISE program is giving me a more balanced college experience as a commuter and transfer student!"

– Jynna Harrell (Biology major)

"I am more of a quiet person, but through REALISE, I have been encouraged to open up and talk with others and that has helped me tremendously. As a REALISE student, I learned to start conversations and follow up with familiar faces I see at each event. It is a great way to reach our goal of creating community within the STEM community."

– Kaitlyn Edwards (Biology major)

Sustainability: As the REALISE grant is coming to a close, we are currently in the process of transforming the PRMs into a college-wide group of student leaders. As faculty mentors, this has been one of our biggest challenges as we work with the PRMs and other key stakeholders (REALISE leadership, college dean, etc.) to decide what form this group should take.

One key lesson that we as faculty mentors have learned is that incentivizing the PRMs for their work is essential. When we were first brainstorming ideas for the program's sustainability, we talked with the PRMs about potential incentives moving forward (wages/stipend, class credit, book vouchers, etc.). Through informal conversations as well as anonymous focus groups, the PRMs decided that turning the program into a student organization would be the best direction. Based on this feedback, the PRMs worked on developing a club constitution and officially became a student organization in Spring 2023. We are now at the end of the Spring 2023 semester and have seen mixed results in the students' commitment to the PRMs now that they are unpaid. In previous semesters, the role of our PRMs required more weekly hours of commitment than most other student clubs and the loss of pay impacted the students' motivation to maintain the same workload. This caused a drop in PRM's engagement in meetings and events which both the faculty mentors and PRMs noticed. In reflection, the PRMs recognized that while the money wasn't their main reason for becoming a part of the program, it did help them prioritize this work and commit more time outside of the weekly meetings. Additionally, we feel that compensating students for their work whenever possible creates the most equitable access to skill-building opportunities such as the PRMs.

Another important facet of sustaining and expanding the program is buy-in and support from our college dean and other leadership. The REALISE "brand" has been strongly associated with the three main departments by many in the college, so having the dean and department chairs involved with rebranding the PRMs as a college-wide group of student leaders will lower the barrier to attaining buy-in from faculty and potential PRMs across departments. Additionally, while some of the PRMs' initiatives are being picked up and funded by our institution's new Quality Enhancement Plan (QEP), the remainder of the event funding and any monetary student incentives will need to be supported elsewhere (Mekolichick, 2023). Buy-in from both the dean's office and department chairs may provide multiple options for funding streams for PRMs, as well as help with recruitment of faculty mentors across more departments. Having a clear, cohesive message and vision for the PRMs college-wide will give this initiative the best chance for success and longevity.

REFERENCES

Clements, T.P., Friedman, K.L., Johnson, H.J, Meier, C.J., Watkins, J., Brockman, A.J, Brame, C.J. (2022). "It made me feel like a bigger part of the STEM community": Incorporation of learning assistants enhances students' sense of belonging in a large introductory biology course. *CBE – Life Sciences Education, 21*(2).https://doi.org/10.1187/cbe.21-09-0287

Estrada, M., Woodcock, A., Hernandez, P. R., & Schultz, P. W. (2011). Toward a model of social influence that explains minority student integration into the scientific community. *Journal of Educational Psychology, 103*(1), 206–222. https://doi.org/10.1037/a0020743

Huston S., Herman R., Liss S., & Taylor B. (2023). Community, curriculum, and CUREs: Transformations in the physics department at Radford University. In J. Briganti, J. Sible, & A. M. Brown (Eds.), *Fostering communities of transformation in STEM higher education: A multi-institutional collection of DEI initiatives* (pp ??-??). Virginia Tech Publishing.

Kennedy, S. A., Balija, A. M., Bibeau, C., Fuhrer, T. J., Huston, L. A., Jackson, M. S., Lane, K. T., Lau, J. K., Liss, S., Monceaux, C. J., Stefaniak, K. R., & Phelps-Durr, T. (2022). Faculty professional development on inclusive pedagogy yields chemistry curriculum transformation, equity awareness, and community. *Journal of Chemical Education, 99*(1), 291-300.https://doi.org/10.1021/acs.jchemed.1c00414

Lane, T.B. (2016). Beyond academic and social integration: Understanding the impact of a STEM enrichment program on the retention and degree attainment of underrepresented students. *CBE – Life Sciences Education, 15*(3), ar39. https://doi.org/10.1187/cbe.16-01-0070

Lau J., Mekolichick J., Raimer A., & Kennedy S. (2023). Assessing changes in student engagement using a mixed-methods approach. In J.Briganti, J. Sible, & A. M. Brown (Eds.), *Fostering communities of transformation in STEM higher education: A multi-institutional collection of DEI initiatives* (pp ??-??). Virginia Tech Publishing.

Mekolichick J. (2023). Institutionally advancing inclusive excellence: Leading from the middle in times of transition. In J. Briganit, J. Sible, & A. M. Brown (Eds.), *Fostering communities of transformation in STEM higher education: A multi-institutional collection of DEI initiatives* (pp ??-??). Virginia Tech Publishing.

Nunn, L.M. (2021). *College belonging: How first-year and first-generation students navigate campus life*. Rutgers University Press.

Stets, J.E., Brenner, P.S., Burke, P.J., Serpe, R.T. (2017). The science identity and entering a science occupation. *Social science research, 64*, 1-14.https://doi.org/10.1016/j.ssresearch.2016.10.016

Zaniewski, A.M., Reinholz, D. (2016). Increasing STEM success: A near-peer mentoring program in the physical sciences. *International Journal of STEM Education, 3*(14), 1-12.https://doi.org/10.1186/s40594-016-0043-2

This program is supported in part by a grant to Virginia Tech from the Howard Hughes Medical Institute through the Inclusive Excellence Grant.

CHAPTER 7.

COMMUNITY, CURRICULUM, AND CURES

Transformations in the Physics Department at Radford University
SHAWN M. HUSTON; SANDRA LISS; BRETT TAYLOR; AND RHETT
HERMAN

ABSTRACT

This book chapter will focus on the efforts made in the Department
of Physics at Radford University to improve diversity, equity and
inclusion (DEI) among physics majors. These efforts include
changes to the curriculum as a whole, changes to individual
classes, faculty training, and attempts to build a welcoming
community for students. Each of these items will be discussed in
some detail. Specifically, an introductory freshman seminar was added;
course-based undergraduate research/problem-based learning was
integrated into classes such as introductory astronomy, atmospheric
physics, geophysics, and thermodynamics & statistical mechanics, as
well as others. In addition, examples will show the ways in which DEI
has been explicitly addressed in the classroom. Historically physics
lacks diversity. While we have much more progress that needs to be
made, we believe our efforts at Radford are one possible step toward
recruiting and graduating a diverse student body.

INTRODUCTION

IDENTIFYING THE CHALLENGE AND EXPLAINING WHO WE ARE AT RADFORD UNIVERSITY.

Physics as a discipline lacks a diverse representation of faculty and students (American Physical Society, 2023). For example, approximately 16% of the college-aged population is African American, and yet only approximately 3% of bachelor's degrees in physics are earned by African American students. This difference lags behind the percentage of degrees earned in engineering and math (approximately 4%), and well behind those earned in chemistry, biology, or computer science (approximately 8%) (IPEDS, Census, et al., 2023). There are undoubtedly many reasons behind the paltry numbers for physics. Identifying these reasons is outside the scope of this chapter, rather we aim to communicate how we have attempted to increase persistence and retention among those students who elect to study physics at Radford University (RU).

Radford is a primarily undergraduate institution (PUI) in rural southwest Virginia. A large population of our students are first generation (Acosta, 2021), and we are a lower cost public school, routinely landing on U.S. News and World Report's best colleges rankings in multiple categories (Brackin, 2021). Radford is not an R1, as such the primary focus of faculty at RU is teaching (approximately 60%), rather than professional contributions (approximately 25%) (Radford University, 2022). The Department of Physics itself is composed of four tenured or tenure-track faculty members, along with an additional full-time instructor (Radford University, 2023). Class sizes are small, with upper-level classes for our physics majors typically being fewer than ten students. Physics majors have faculty members for multiple classes throughout their academic career. These factors allow us to build relationships with our students, to attempt to build a community of learners, and to help students identify as scientists.

LEVERAGING RADFORD'S STRENGTHS

It is undeniable that many first-year college students are overwhelmed by the challenges in their new environment. This feeling can particularly be the case for first generation students. The university experience as a whole can be daunting, let alone classes themselves. An isolated student, faced with new academic challenges may be left wondering

Figure 7.1: A small classroom

whether they truly belong in their major or in college itself. Identifying this problem, isolation and a lack of a sense of belonging, is an important step to addressing retention issues in general and in physics specifically. The best thing we can do for our students is offer them support and create personal connections for those students, with each other, and with ourselves. One of the ways in which we have facilitated a network of support is by having dedicated spaces for our physics majors to congregate and having a strong physics club presence.

Radford's focus on undergraduate education and small class sizes allows our physics faculty members to develop years-long relationships with students. As such, we can talk about issues like imposter syndrome with some level of credibility. Radford's small class sizes also allow us as faculty members to step outside some of the traditional class structure. A "normal" physics class is composed of homework, tests, and possibly labs. With large class sizes this is about all that a single faculty member can handle. We have been able to step outside of those bounds and introduce course-based undergraduate research and problem-based learning into many of our classes. This kind of pedagogy builds marketable skills for our students and helps them identify as scientists because they are, in fact, doing science.

OPERATIONAL STEPS TO ACHIEVING THE GOAL: INCREASING DIVERSITY IN PHYSICS

FACULTY EDUCATION

It is an unfortunate human truth that most of us will initially react defensively or in denial when it is brought to our attention that we have an area that we need to improve on. For this specific problem, lack of diversity in physics, that might look something like this: "Okay, I recognize that the numbers for physics don't look good, but what does that have to do with me? I'm not racist. I'm not doing anything wrong. What does teaching physics have to do with race anyway? Physics is just physics." That certainly was the first author's thoughts when first confronted with the issue. The first barrier in establishing change in education is achieving faculty buy-in. For physics, buy-in should be relatively simple, a glance at the statistics and a few moments of thought is all it really takes to identify that surely something must be done in order make the physics community more equitable. Once the problem is identified, diversity training is extremely helpful. For instance, the first author had never heard of stereotype threat and had little understanding of microaggressions before taking part in an extended sequence of diversity training sessions offered through REALising Inclusive Science Excellence (REALISE), a program here at Radford supported by a Howard Hughes Medical Institute (HHMI) grant. In fact, the first author's approach to talking about race in the classroom was to never, ever do it.

Great, faculty buy-in has been achieved. Now what? After acknowledging that we needed to go beyond just teaching the subject, a number of steps were and are being taken to provide an equitable and inclusive experience to recruit and graduate a diverse student body here in RU's Department of Physics. Our focus is on increasing persistence and retention in the major for all students, but particularly for traditionally underrepresented students, which, in physics, includes women (IPEDS & APS, 2023).

ESTABLISHING COMMUNITY AS A DEPARTMENT

One way of increasing student retention is by increasing a student's sense of belonging as well as student access to support. We were fortunate to be the beneficiaries of a renovation to our existing science building just a few years ago. While we had limited control over the building spaces as a whole, we made it a priority that our physics majors have a space set aside for their own use. This comes in the form of a dedicated club space for our local chapter of the Society of Physics Students. We viewed this space as such a high priority that we cut this space out of our multi-use faculty research lab. We were also fortunate to have a large study area (pictured above) that is immediately outside of our department faculty offices. During most hours of the day our physics majors can be found congregating in this space, which is a great area to study, do homework, and talk to one another, and which allows them ready access to faculty members, many of whom have an open-door policy and/or stop by that space to chat with students multiple times a day. It is our hope that this unique environment makes students more likely to see faculty members as approachable. We certainly emphasize that we are there for them, and we view our students as whole people, as we hope they view us.

PROGRAMMATIC CHANGES

Freshman Seminar (PHYS 201)

PHYS 201 was introduced as a required course for all physics majors starting in Fall 2020. It is a one hour seminar course that is all about building community and improving students' identity as young physicists. In this course a great deal of time is spent talking about these topics and providing external resources for students in an attempt to build that identity. These external resources are composed of both groups and individuals with many of these focused on the under-represented minority students in the group, providing links to external bodies/people that are not like those who traditionally make up the physics community (predominantly white and male). An example of one of these external bodies is the National Society of Black Physicists. The other large faculty effort in establishing PHYS 201 was assembling an alumni panel to speak with our freshmen. The panel participants are composed of the most diverse group of RU physics alumni that could be assembled. They are not all male and white and include a member of the LGBTQ+ community, multiple non-white participants, and about 30% of the participants are female. This alumni panel is hugely popular with our students. It allows them to talk with people who were once in their shoes, are diverse, and are successful in a broad range of post-baccalaureate careers. PHYS 201 also includes an introduction to possible career paths. In the discussion of careers, we have also tried to provide examples of physicists who are not white/male. We also have had the REALISE students, a group of peer mentors at RU, present a peer-level discussion about diversity/belonging/identity issues (see the chapter "Assessing Changes in Student Engagement Using a Mixed-Methods Approach" in this book). **Since adding this class that specifically targets community building and scientific identity, retention numbers for students moving into their sophomore year as physics majors have increased.** While retention has increased, it must be admitted that due to the typical cohort sizes of our freshmen majors (approximately eighteen to twenty enter each year), the sample size is small.

Introducing problem-based learning (PBL) projects into classes

We base so much of our grades and assessments of our students' work on items that they will NOT encounter when they are employed or actually doing their graduate school research. By changing the classroom experience, we can change student engagement and enhance retention. Jobs and graduate research projects do not have tests, homework, and labs as we use them – correctly, it might be added – in most of our instruction. Also, most students, if they just put their traditional coursework and GPAs on their resumes, would present essentially identically to potential employers or graduate schools. Our faculty have received a number of emails from our alumni with a common theme, saying something along the lines of "While I never did so well on [physics] tests, I'm now doing really well in [their current career] and am finally using what I learned in class." That common phrase of "...finally using what I learned in class...." really stands out. In other words, while they *learned* something in class, they did not *use* that knowledge while they were undergraduate students with us.

There are multiple ways to assess student learning in addition to the canonical test and homework grades. One of these is through a grade in problem-based learning (PBL), either embedded within a class or even as a class by itself. While we will present some ways in which PBL has been embedded into individual courses, we offer the following observations first. Students tend to do very well in the PBL aspects of the class. We have often observed that the students put proportionately more time and effort into these projects than in the traditional aspects of the class (homework and tests). They have frequently commented that they are quite proud of the work because they can see actual results. They say that it feels more like a real job or graduate school type of project, with a well-defined goal, or "deliverable" to be completed.

The Department of Physics at Radford offers a special Arctic Geophysics research class every other year. Over the past decade, the faculty member who teaches that class, Dr. Rhett Herman, has been slowly moving away from the traditional test/homework/lab grading scheme for that course into a more research-oriented format. This class is, unfortunately, an isolated elective class that affects a small number of students. In his second year of working with our REALISE/HHMI program, Rhett encountered the more formal idea of Project-Based Learning (PBL), and an idea was hatched to include PBL in not just the Arctic Geophysics class, but in other (required) classes as well, so that this type of learning would be available to all of our physics majors. The PBL process was introduced in a two-day workshop facilitated by members of the Center for Project-Based Learning at the Worcester Polytechnic Institute (WPI). They detailed how to incorporate real-world projects into classes, and how those projects could meet one or part of a learning outcome for a class. This incorporation would mean that the learning outcome (or part of one) would be delivered to the students, and assessed by the instructor, through that project. Further, this learning outcome would not be discussed in the lecture or even included on the in-class exams. Two examples of learning outcomes that could be addressed through PBL are (1) Students will apply physics to real-world problems, and (2) Students will effectively communicate science to a general audience. Others follow.

These projects are sometimes difficult to create. However, they are rewarding for both the students and the faculty. One of the main criteria that is used for these projects – in addition to making sure they address a learning outcome – is for the project to answer the following question in the affirmative: "Is this project something that can be listed as a separate line on students' resumes?" Will this project result in students learning something, or developing some skill, that they can put on a separate line on their resume? The PBL process would include skills such as additional programming, using electronics to build data-acquisition sensors, developing certain experimental skills, or communicating scientific results to the public.

The PBL process has been completed in a number of upper-level classes and has been started in an introductory class. A short list of some of the upper-level classes and examples of PBLs that have been used in these courses follows. In all cases PBL is just one aspect of the course, except the explicit example of the Arctic Geophysics class previously mentioned. Note that anything related to sensors has employed Arduino microcontrollers. These all involve basic electronics as well as programming, in addition to the inevitable data analysis and end-of-semester public presentations. Most of these involve some small engineering skill in order to house/hold these sensors.

PHYS 330 – Thermodynamics and Statistical Mechanics

- Develop micro-climate sensors that could monitor some thermal aspect of a campus building that relates to the official measured energy usage of that building.

- Develop thermal sensors that can monitor the heat absorbed and re-emitted from asphalt parking spaces, with some the standard black color while others are painted white.

- Develop irradiance sensors that can monitor the percentage of infrared radiation in the solar input and reflected/re-radiated output from various ground cover surfaces that could be on campus.

PHYS 301 – Atmospheric Physics

- Develop micro-climate sensors that could monitor some aspect of the area around a stream bed that is to be re-established along its natural route.

PHYS/GEOL 406 – Geophysics

- Develop a survey plan and carry out a geophysical site survey at a nearby undeveloped industrial park using our professional-level geophysics equipment. The goal is to determine the fitness of these large (approximately 20 acre) parcels for siting a large (>100k sq. ft.) industrial building. The final report is delivered to the controlling multi-governmental agency to use in marketing these sites to potential buyers.

- This field project has been used in the 4 most recent offerings of this every-other-year (spring) class.

- Alumni have reported that they have been excited to learn how their work has helped companies determine whether to site their buildings in that area.

The above are all examples of PBL embedded within a course. Each of these projects counted toward approximately 10% of the overall class grade, and students have, throughout the years, relayed that they have put many of these projects in as separate lines on their resumes, along with the skills that they have used to complete these projects. They have also shared that the end-of-the-semester presentations required by these projects – the "deliverables" – have been topics of discussion in job and graduate school interviews

PHYS 324/325/326 – the Arctic Geophysics Research Experience cycle of courses

The Arctic Geophysics Research Experience that is taught every other year is an example of extensive use of PBL. This is a one-year sequence of two classes, the one-credit-hour PHYS 324 – Arctic Geophysics Preparatory Seminar and the four-hour PHYS 325 – Arctic Geophysics Field Research. PHYS 324 is offered in the fall of odd-numbered years and PHYS 325 is in the subsequent Spring semester (even-numbered years). Each of these is 100% PBL and is entirely student-centered. There is also a third class offered in this experience in the subsequent Fall, the one-hour PHYS 326 – Arctic Geophysics Capstone Seminar. Students are not required to take this course because a number of those enrolled in the research experience are seniors who graduate after taking PHYS 325. However, most who have not graduated take the PHYS 326 class. The Arctic Research Experience is open to any major, although most of the students in the past have majored in Physics or Geology (typically with a Physics minor). Chemistry, Biology, Math, and Computer Science majors have also successfully completed this research experience.

In PHYS 324, students develop their own ideas for research projects that quantify some aspect of the arctic sea ice or the arctic environment. They are encouraged to pick this research project with an eye toward their future career path. They could choose something that directly relates to their chosen career either as a direct research topic or through having them learn a particular skill that they can use to enhance their resumes. However, their project could also be something that's not directly related to their career, but something that interests them as a research topic.

After students pick their research topic, they write their own research proposals. These proposals must involve some quantitative aspect of the artic environment and thus will necessarily involve some type of sensor. These projects are typically designed using either Arduino microcontrollers or Raspberry Pi microprocessors due to both their inexpensive and Do-It-Yourself (DIY) nature. In addition, there are a great number of inexpensive sensors designed to work with these items that are the same as the "professional grade," very expensive, black-box sensor packages that are commercially available. However, we emphasize to students that they do not need to know **anything** about these items beforehand. The faculty member's role as mentor is to show the students what they need to know in order to bring their project to fruition. Their job is to identify a sensor that will address their research question, and then work closely with the faculty member to come up with a plan to both build the sensor and carry out the research. Thus, they have to come up with a reasonable budget for their project, again learning something that they will have to know later for a job, a graduate school proposal, or a research proposal.

After the students turn in their proposals, identifying information is removed, and the members of the class serve as the "review committee of the whole" for everyone's proposal (including their own). After receiving the feedback (anonymously), the students submit their revised proposal for (reasonable) funding by the Department of Physics. The necessary equipment is then gathered, and the faculty member works closely with them as they start to build their sensor packages. These sensor packages must be made into self-contained units with batteries, microSD cards for the data, switches, and everything else required of the more expensive, commercial sensors that they will use later in their careers. Students usually get a good start on their builds during the Fall semester, and they start to learn the difficulty in bringing even a small research project to fruition starting from absolutely nothing.

In the Spring class, PHYS 325, students spend the first half of the semester finishing their builds and testing them locally. They learn to troubleshoot and fix issues that their testing reveals. This troubleshooting is often both frustrating and very rewarding, giving students an appreciation for a non-textbook-perfect research project. The students do have to work surprisingly fast under the deadline of traveling to Alaska at the end of the sixth class week to deploy their sensors on the ice. They travel to Alaska during one of two weeks (half the class each week) in late February to early March, when the sea ice is thickest. They are again learning to troubleshoot under difficult conditions of extreme cold (although we do stay in warm buildings – we are not camping in the arctic!), and not having the usual resources of the Radford Physics laboratories. They had to make sure to take all of the supplies and tools that they anticipate using when things inevitably break, or go wrong, during their deployments. They must also learn to rely on each other because most of the projects require some helping hands to make them happen.

After the trip, the students present their results to the University community at a campus-wide research forum in mid-April. One of the main aspects of PBL is the public communication of their work. They find that this presentation helps them truly understand what they have accomplished in this course series.

In the subsequent Fall, most (if not all) of the Alaska students who have not graduated take the one-credit-hour PHYS 326 – Arctic Geophysics Capstone Seminar. In this class, students further analyze their data from the Alaska research trip and prepare a professional presentation. So far, all of these have been at the Fall Meeting of the American Geophysical Union, the world's largest gathering of earth and space scientists (approximately twenty-five thousand attendees from all seven continents). The most recent cohort saw four of the non-graduating students take PHYS 326, with one who had just graduated "participating" in the same professional preparation remotely, just as if she were enrolled in the class.

The 2023 AGU Fall Meeting saw the largest Radford University contingent ever, with five students attending, two of whom were female and one a non-majority male. These five students had a total of 4 poster presentations, spread out over three separate sessions on different days. **Two of the students were offered direct graduate school positions**, although those graduate offers were not in the career paths that the students preferred. This offer speaks to one of the main goals of the entire Arctic Geophysics experience – to get students the full research experience that they need in order to advance their careers in whatever direction they choose. This project truly is student-centered and student-driven.

Students self-identify as scientists by being scientists and undertaking scientific studies themselves

In the previous section we offered extensive examples of changes made to courses to incorporate PBL. These examples all fit under the umbrella of students self-identifying as scientists. Additional efforts have been made to enhance student self-identity in the Introductory Physics sequence for majors, PHYS 221 and 222, as well as our pair of skill-building classes for majors in their sophomore year, Mathematical Methods in Physics (PHYS 303) and Computational Methods in Physics (PHYS 370).

Students in the introductory sequence often complain about, and we joke about, no friction, no air resistance, etc. Of course, the reason we start with these simplifying assumptions is because the mathematics to solve such situations analytically are typically beyond the students' current mathematical skills. A large effort has been made in these two courses to work on building identity as physicists for students and their ability to solve "non-trivial" problems by using computational physics. The hope is that students can see themselves more as physicists because they are not limited to the simplest situations and can include the physics that they see around them without being limited by their current mathematical skills. This kind of inclusion is also particularly important in the second semester class because students have difficulty visualizing topics in electricity & magnetism, whether that's field, electric potential maps, motion under electric and/or magnetic forces together, or assorted other topics. They really have to think three-dimensionally, but if they can build their own system computationally and see directly what it looks like, then the hope is they will see themselves as more accomplished than they might without those skills.

Our sophomore skill-building classes, PHYS 303 and 370, are supposed to again build a sense of identity and a recognition of students' own skills and their ability to solve physics problems. Mathematical Methods in Physics, PHYS 303, has been incredibly helpful in building student confidence, and students, including a math major who graduated a few years ago, have told us it was the most useful skills course they took overall. The computational course again builds students' skills and requires them to complete a final self-chosen project. The hope is to build their skills (both in computation and in presentation both written and oral) and help them see themselves as REAL physicists who can solve non-trivial problems.

Further examples of helping students self-identify as scientists come from our sequence of astronomy courses, in which students take on aspects of astronomy research projects, such as applying for observational time on remote controlled telescopes, planning observational runs, and processing data themselves. These activities are all incorporated into regular classes and while students may opt to continue their work through independent study or guided undergraduate research, their initial "taste" of this research comes as required coursework. This process has been extremely well received by our physics and astrophysics majors.

Addressing imposter syndrome and diversity in physics – why are "all" these historic physicists white men?

If one pops open any textbook on the development of physics and studies the people behind that development, one finds an unsettling trend. Nearly all of these historical figures, and certainly all of the well-known ones, are white men. As mentioned in the introduction, a lack of diversity in physics is still extremely prevalent. As much as we have tried to increase diversity in our classrooms in the Department of Physics at Radford, the same trend exists here. If you look in any of our classes, you will see that the majority of the students are white men. It is essential that we increase the retention of non-majority and female students to reverse this trend. In the sections above, we have outlined some ways in which we have attempted to increase retention among all groups of students. These efforts are further enhanced by addressing imposter syndrome and diversity in physics directly and explicitly in our classrooms.

Explicit discussions of diversity in physics take place in our Freshman Seminar, in our Introductory Astronomy courses, and in Modern Physics, a third semester class for our majors which is heavily focused on historical development of physics in the early 1900's. The Astronomy and Modern Physics classes are particularly suited to these discussions because they not only deal with the problem-solving side of astronomy and physics but the history behind it as well. This focus allows for organic discussions of historical examples of privilege; one illustration of this is the number of Nobel Prize winners whose parent(s) were also Nobel Prize winning scientists. We offer no condemnation of these scientists or their work, but we point out the advantages they had. We also address historical discrimination against women and non-majority persons in the development of astronomy and physics. In our Introductory Astronomy class for majors, students give a presentation on a scientist who interests them. Students are encouraged, but not required, to select non-majority or female scientists. For those who have chosen non-majority or female scientists, it is often clear during their presentations that the representation seen is impactful to them.

Discussions about imposter syndrome are also an important component of helping students understand that they do, in fact, belong in the major. This discussion requires real vulnerability on the part of faculty members and alumni, in the case of the alumni panels in the introductory seminar. By explaining to the students that at some moment in time everyone in the class is likely to feel like they just aren't smart enough for this major or that they don't measure up to their peers, we increase the chance of these students persisting. Faculty members regularly explain to students that they also felt that way themselves at times and that we have alumni from Radford who we taught, who we know felt that way, who have gone on to fantastic careers as scientists. Several times after these discussions, we have had students approach us and offer deep thanks for addressing this topic. Imposter syndrome is rampant in physics, and for those students who are already in the minority in the classroom, it can be a strong detractor toward completion.

All of these discussions, whether they are about privilege, discrimination, or imposter syndrome require real authenticity from the faculty member leading the discussion. This action goes well beyond checking off a DEI box.

Additional changes that may help some students to overcome boundaries

We offer a few further comments on ways we have helped students overcome boundaries. Where appropriate we have included lectures on metacognition and reflective corrections in our classes to help students understand how to think deeply. Cost is often a barrier for non-majority students, so we have switched textbooks to open resource or low-cost alternatives when we can. In one instance, we have even offered testing to an English as a Second Language (ESL) student in his native language, though care must be taken if doing this to make sure testing is equitable. Last, homework and test problems have been chosen to be more gender neutral. One of the first physics problems a student typically encounters is "A boy throws a ball..." because these are relatively easy to solve. This is great, but potentially not very interesting for (a) people who are not boys or (b) people who don't particularly care for sports.

WHAT WE HAVE LEARNED

First, our department's changes are an on-going, collaborative effort. Measuring progress is difficult with small numbers, and the effects may not be immediately seen. We are collecting long-term retention numbers as a measure of success. Although small sample sizes can be deceiving, we agree that making a difference to one or a few female or non-majority physics students equates to success. It may have also been possible to measure change by taking surveys of students' perceptions pre- and post- HHMI initiative. This would have required some forethought, because we would have needed to survey students from approximately 2019 so that we could compare them to today's results. These surveys would, of course, be clouded by the impacts of the dramatic changes in emotional well-being caused by recent events, e.g. COVID, that are evident broadly in today's student body. Although we did not have the foresight to survey our students pre- and post-HHMI, it may be possible to develop a pre- and post-course or first year and graduate surveys to measure change in sense of belonging and science identity (Potvin & Hazari, 2013). This is a step that may be undertaken in the future.

Second, integrating PBL is a lot of work, and course-based undergraduate research experiences (CUREs) are not easy. Projects fail. Adding CUREs/PBL to traditional classes means that some of the traditional material must come out. The educational impact is, however, worth it. Part of this reward can be seen in the pride the students take from their results and the independence that these projects foster.

Thanks for reading.

REFERENCES

Acosta, M. (2021, November 5). *Center offers support, advocacy for first-generation students*. Radford University. https://www.radford.edu/content/radfordcore/home/news/releases/2021/november/first-generation-students.html

American Physical Society (2023). *Minority physics statistics*. [Data set]. American Physical Society. https://www.aps.org/programs/minorities/resources/statistics.cfm

Brackin, B. (2021, September 13). *Radford University improves and adds ranking in U.S. News and World Report's Best Colleges listing*. Radford University. https://www.radford.edu/content/radfordcore/home/news/releases/2021/september/us-news-improves-ranking.html

Integrated Postsecondary Education Data System (IPEDS), & American Physical Society (APS). (2023). *Physics degrees earned by women*. [Graph]. American Physical Society. https://www.aps.org/programs/education/statistics/womenphysics.cfm

Integrated Postsecondary Education Data System (IPEDS), United States Census Bureau (Census), & American Physical Society (APS). (2023). *Bachelor's degrees earned by African Americans by major*. [Graph]. American Physical Society. https://www.aps.org/programs/education/statistics/aamajors.cfm

Lau J., Mekolichick J., Raimer A., & Kennedy S. (2023). Assessing changes in student engagement using a mixed-methods approach. In J.Briganti, J. Sible, & A. M. Brown (Eds.), Fostering communities of transformation in STEM higher education: A multi-institutional collection of DEI initiatives (pp 49-63). Virginia Tech Publishing.

Potvin, G., & Hazari, Z. (2013, July 17-18). *The development and measurement of identity across the physical sciences*. Physics Education Research Conference 2013, Portland, OR, from https://www.compadre.org/Repository/document/ServeFile.cfm?ID=13182&DocID=3729

Radford University. (2022, December 2). *Radford University teaching and research faculty handbook*. https://www.radford.edu/content/dam/

departments/administrative/faculty-senate/
Teaching_and_Research_Faculty_Handbook-12-2-2022.pdf

Radford University. (2023). *Physics and astronomy.* Artis College of Science and Technology. https://www.radford.edu/content/csat/home/physics.html

---

This program is supported in part by a grant to Virginia Tech from the Howard Hughes Medical Institute through the Inclusive Excellence Grant.

CHAPTER 8.

WE'RE ALL IN THIS TOGETHER

Fostering a sense of community within a biochemistry department
ANNE M. BROWN AND SASHA C. MARINE

INTRODUCTION

From in-person symposiums to virtual chatrooms, we have strived to instill a sense of community, belonging, and agency for our students, faculty, and staff within the Department of Biochemistry at Virginia Tech. Biochemistry is a challenging major, with many students wanting to attend a wide variety of professional (e.g., medical, dental, pharmacy, veterinary, etc.) and graduate programs after graduation. We are well-known (for better, or worse) for a six-credit laboratory applications course taken by students in their junior or senior year. This course pushes students in performing, writing, and presenting their scientific experiments in numerous lab reports and a practical exam. Until the late 2010's, we often did not have substantial curriculum interaction with our in-major students until their junior (third) year, as courses such as General Chemistry, Biology, and Calculus are taught by other departments. With curriculum changes underway and by participating in the Inclusive Excellence program, we sought to transform the learning and community environments of our students, staff, and faculty.

In the literature, some of the biggest challenges faced by historically marginalized communities are unrealized and unaddressed in science, technology, engineering, and mathematics (STEM) fields, contributing to lower graduation rates (Xu et al., 2018). However, studies suggest that both learning communities and communities of practice have significant impact on improving success, perseverance, and sense of belonging in populations that are historically underserved or underprepared for collegiate level coursework. Engaging students both academically and socially in community/communities has been shown to improve student outcomes and persistence (Gopalan et al., 2019). Taken together, we recognized the need for consistent means of engagement outside the traditional classroom and the potential impact it could have on students, especially those from historically marginalized communities.

A group of motivated faculty endeavored to use our initial year in the Howard Hughes Medical Institute (HHMI) Inclusive Excellence program to build knowledge and determine what might be the best fit for activities and transformation within our department to support all members of our community – students, staff, post-docs, and faculty. Questions we asked ourselves as a small group wanting to make big impacts focused on topics such as: What can we do to increase a sense of belonging and community for students in our major? What tools and knowledge should we emphasize for faculty to enhance their knowledge base on inclusive excellence? We started out with small faculty group meetings, occurring monthly to discuss these avenues. Each month, the group had a paper for discussion. Growth mindset – especially the work by Canning et al. (2019) was a huge motivator in our planning of several avenues to increase the sense of community and belonging within our department. Inspiring students and celebrating them became a common thread of discussion and the need for that in our department was apparent. We next focused on what elements would be most valuable to all members of our community – students, faculty, postdocs, and staff – and how we could establish initiatives that were both sustainable and focused on supporting the needs of each of these groups.

INITIATIVE #1: A POSTER SYMPOSIUM

When you think of a poster symposium, what comes to mind? Likely you picture a crowded room with stationary presenters and poster stands affixed with printed posters. You may think of poster judges, who evaluate the competing entrants, or you may think of sponsor tables and their promotional giveaway items. Because people often have a preference for the familiar (i.e., the familiarity principle; Zajonc, 1968) and established (i.e., status quo bias; Samuelson & Zeckhauser, 1988), it should come as no surprise that we initially modeled our departmental poster symposium after the poster sessions we had attended in the past. We focused on this initiative because 26% of our in-major students conduct undergraduate research every semester, either for course credit (15%) or as part of a course (i.e., course-based undergraduate research experience or CURE; 11%). We named the event Engelpalooza, as this amalgamation honors both our departmental founder (Charlie Engel) and the impressive contributions of our students (Merriam-Webster Dictionary).

IMPLEMENTATION OF ENGELPALOOZA

Our inaugural symposium was held indoors on a weekday in October in 2019. It had posters displayed along the hallways of the departmental home building on campus, with a free, boxed lunch set up in an adjoining auditorium-style classroom. The event was heavily advertised to our undergraduate community (N=471), and thirty-four undergraduate researchers presented posters in two sessions at the event. We did not have any graduate student presenters, although the Biochemistry Graduate Student Assembly (BcGSA) did participate in the event via an info booth and by giving out ice cream. The department's Advisory Board served as poster judges, and attendees could enter a raffle to win departmental swag. More than a hundred students, faculty, staff, and alumni attended the inaugural poster symposium.

INAUGURAL POSTER SYMPOSIUM: FEEDBACK

Following the event, presenters were asked to provide feedback in a short electronic survey. Our response rate for the survey was 32% (11 of 34). Comments were generally positive, with 70% of respondents indicating the event was very beneficial, and 60% of respondents indicating they were very likely to recommend the event to a friend or classmate. The department's Advisory Board also had a favorable opinion, describing the event as a strong community-building experience. In terms of areas for improvement, survey respondents requested more networking opportunities, better event marketing and communication, and a less crowded venue. The faculty organizing the event felt it was important to take additional actions to promote inclusion, namely, to solicit graduate student presenters and to discontinue the poster competition.

SUBSEQUENT POSTER SYMPOSIA: MODIFICATIONS AND FEEDBACK

Following the inaugural poster symposium, we have worked to incorporate the suggestions. Starting in 2020, for example, we have advertised the event equally to our undergraduate (N=471) and graduate (N=43) communities. This has resulted in better parity, with approximately 70% of posters being presented by undergraduate students and 30% of posters being presented by graduate students. The distribution somewhat reflects the difference in cohort size though a greater proportion of graduate students participate in the event than undergraduate students. We also consistently have between twenty and forty research poster presenters each year. To facilitate networking, we have info booths for both the Biochemistry Club (an undergraduate organization) and the BcGSA (a graduate organization) at the event. We also invite an accomplished alumnus to give a keynote presentation and attend the poster sessions. In terms of marketing, we have increased how often we are advertising on the departmental website, Facebook, and Twitter accounts in the weeks leading up to the event. We have also developed a promotional video, paid to have the university mascot make an appearance at the event, and purchased photo backdrops and banners.

The poster symposium continues to be held on a weekday in October, but starting in 2021, we have held the event outside under a tent on a lawn near the departmental home building. This less crowded venue enabled inclusion of additional info booths on ASBMB degree certification, study abroad opportunities, undergraduate research 101, and our department's peer mentor program. It also gave us space for live demonstrations on nanopore sequencing, robotic liquid handling, and virtual reality simulations of molecular dynamics. Another benefit of holding the event outdoors is increased attendance – we now have more than 170 students, faculty, staff, and alumni come to the poster symposium. We continue to provide a free lunch but have expanded the options to better accommodate dietary restrictions and food allergies. We are proud the event allows us to authentically, and sustainably, foster a sense of community within our department.

Following the 2021 poster symposium, presenters and attendees were asked to provide feedback in a short electronic survey. Our response rate for the survey was 42% (49 of 117). As before, comments were generally positive, with 94% of respondents indicating the event fostered a sense of community. We were also pleased to learn that 90% of respondents were satisfied with the amount of communication leading up to the event, and 83% of respondents were satisfied with both the outdoor venue and with the networking opportunities at the event. In terms of areas for improvement, survey respondents requested more options for displaying diverse poster types (such as vertical or electronic posters) and more involvement from departmental personnel.

SPECIAL COMMENTS: HOSTING A VIRTUAL POSTER SYMPOSIUM DURING A PANDEMIC

In the fall of 2020, COVID-19 guidance and the public health situation precluded an in-person poster symposium. However, we were aware that the pandemic was increasing anxiety (Chrikov et al., 2020) and loneliness (Giovenco et al., 2022) amongst college students, with individuals identifying as an under-represented minority being more likely to experience anxiety and depression (Soria et al., 2020). To provide an opportunity for safe social connection and scientific networking, we decided to hold our poster symposium online using a Discord server. Discord is a free, user-friendly platform that enables individuals to set up topic-specific channels (i.e., servers) for collaboration, communication, etc. (Lacher & Biehl, 2018). A total of 22 presenters – 10 undergraduate and 12 graduate – shared electronic posters with an audience of more than 140 students, faculty, staff and alumni. We also had a panel discussion with current undergraduate researchers, who answered audience questions about how to get involved in research, what lab culture is, and how to list lab skills on a resume. To facilitate community building, we held a pet photo contest with winners decided by popular vote. We also had a Discord channel open for non-science discussion for the duration of the event. Overall, participants had a favorable opinion of our virtual poster symposium, describing the event as worthwhile and engaging.

> **Considerations when developing a poster symposium for community building**

While it may seem obvious, the goal of the event should influence how it is designed and implemented. Small, structured events better facilitate networking and interpersonal development (Mowreader, 2023), while having fewer posters increases attendee circulation and event flexibility (Rowe and Ilic, 2015). We choose to keep our event modest in scale because we believe an intimate learning environment is valuable to our student community. When developing your own poster symposium, be mindful of the composition of the organizing committee. Research has also shown that increased diversity within the organizing committee is associated with improved conference speaker gender parity (Casadevall and Handelsman, 2014; Sardelis and Drew, 2016). Our inaugural poster symposium was designed by a small group of tenure and non-tenure track faculty, with a similar ratio of men to women. Subsequent symposia have been organized by a larger committee of faculty, staff, and graduate students. Diversifying the administrative side of our event resulted in better decision-making and a more inclusive poster symposium. To encourage undergraduate attendance at your event, you may consider offering opportunities for students to practice professional social skills before the event (Flaherty et al., 2018), which can reduce anxiety associated with the unfamiliar setting, or having student-led sessions (Pedersen et al., 2013), which can foster science identity and academic self-concept. In our poster symposium, for example, we have had student-led panel discussions and live demonstrations. Faculty are often present as attendees, but rarely moderate or interject.

INITIATIVE #2: AN INSTRUCTIONAL TOOLBOX

We developed a sharable "instructional toolbox" to foster a more inclusive academic environment in our department. The idea came from conversations at faculty meetings, in which colleagues expressed interest in learning how to apply diversity, equity, and inclusion (DEI) practices into science undergraduate courses, but did not know how to get started (hence, the toolbox)

DESIGN AND IMPLEMENTATION

The toolbox was designed as an online repository of research articles, professional development courses, and educational videos, all hosted within our university's learning management system (LMS). A small group of tenure and non-tenure track faculty led the effort, but all faculty in the department were encouraged to contribute to the process. By providing data-driven resources to instructional faculty and graduate teaching assistants, we had hoped the toolbox would broaden users' perspectives and empower them to set goals with specific outcomes to develop more inclusive classrooms. Developing the toolbox took several months, and the online resource was officially launched in mid-February 2020.

TOOLBOX ADOPTION

Invitations to the LMS page were sent to all faculty and graduate students within our department (N = 54), and the toolbox was promoted in a department-wide email and at a faculty meeting. However, less than a quarter of those receiving an invitation to the LMS page accessed it within the first week (8 of 54). To encourage toolbox utilization, a small financial incentive was offered to users who submitted an inclusive teaching rubric, participated in a DEI activity or event, and submitted a reflection post on the discussion board within the Spring semester. Not a single faculty member or graduate student pursued the financial incentive, and by mid-March, our university had shifted to fully remote instruction because of the pandemic. In the subsequent months, we added resources to the toolbox about teaching online, fostering community despite social distancing, and preserving equity in turbulent times, but utilization of the toolbox remained low that spring semester, with zero page views for 10 of 13 weeks.

Disappointedly, engagement with this sharable resource never improved, despite our best efforts. Invitations were re-sent in 2022 and 2023 to the original list of prospective participants, as well as to 22 other members of our department (including staff and postdocs), but 18% (14 of 76) of invitations remain unaccepted. Of the participants who accepted the invitation, 35% (22 of 62) have never accessed the page, and 48% (30 of 62) have spent less than 10 minutes on the page. We have continued to offer the financial incentive for toolbox utilization every semester since 2020 but have never had anyone participate. The toolbox is periodically mentioned in faculty meetings, but reminding colleagues about this resource has not translated into meaningful adoption by users. In contrast to the success of our poster symposium, this route to build community within our department has largely failed.

Considerations when developing an instructional toolbox for classroom inclusion

We had not anticipated how challenging it would be to get departmental personnel to utilize our instructional toolbox. However, conversations amongst the faculty involved in the resource's creation have identified several potential barriers to its widespread adoption. To start, the toolbox was launched partway through the semester, and shortly before COVID-19 drastically altered everyone's professional and personal lives. The timing of this release, coupled with the increase in student-directed emotional labor (which involves actively monitoring and catering to students' emotional needs, often at the behest of one's own) (White Berheide et al., 2022) and the decrease in work productivity (Esquivel et al., 2023) during the pandemic, likely explain the lack of engagement with the toolbox in spring 2020. To capture interest and encourage timely utilization of your sharable resource, we recommend launching it in the weeks prior to the start of a semester, when instructional faculty and graduate teaching assistants are actively preparing course materials. Begin with a minimal approach, carefully curating what research articles, educational videos, etc. you include, to avoid choice overload / paralysis amongst participants (Reutskaja et al., 2020). One may also find it worthwhile to conduct a survey to identify what people want to learn about inclusive academic environments and what resources they think would be helpful. We believe we included too many resources in the initial iteration of the toolbox, and by not delineating the resources by topic of interest, potential users to the LMS page were overwhelmed.

Another potential barrier to our toolbox's widespread adoption was lack of incentivization for DEI within academia. Research has shown that faculty are less likely to engage undergraduates in research (Eagan et al., 2011), adopt open educational resources (Todorinova & Wilkinson, 2020), or engage in pedagogical innovation (Brownell & Tanner, 2012), when a reward structure is lacking. At our institution, expectations regarding DEI have only recently been added to the departmental promotion and tenure process. Faculty participation in DEI activities can be facilitated via expanded and more flexible measures of impact (O'Meara, 2022) and via parsing down of other service obligations (Sullivan, 2021), but few academic institutions do both. Consider how your department can incentivize instructional faculty and graduate teaching assistants to meaningfully adopt DEI practices into their classrooms. We believe our financial incentive of $100 – $250 was too low to entice departmental personnel with heavy workloads to engage in additional activities. We may have had better adoption if we had offered individuals the option to replace a service commitment for a semester with the toolbox (i.e., engaging with materials to broaden their perspectives, implementing DEI practices to foster more inclusive classrooms, and reflecting on whether they were able to achieve their instructional goals).

REFLECTIONS

We sought to improve the inclusivity in our department with the following initiatives: a student-focused celebration and research symposium and a toolbox of easy-to-access learning materials, academic papers and educational videos for faculty to improve their knowledge base. We initially thought these events and resources, as standalones, would improve the community naturally in our department. What we learned was that authenticity and fully integrating student feedback, and breaking down barriers for student voices to be heard would be the most meaningful and true community building aspects of our department. With a pandemic occurring right after our first inaugural year for a student research celebration, it took a lot to keep engaging our students and faculty in what was a very challenging few years. In reflection, we have found success in uplifting student opportunities and accomplishments, and that has gone a long way toward creating true community among students, staff, and faculty. What has become the most noticeable is the increased student participation and engagement in all avenues of the department since implementation of our work. Students have reinvigorated the long stagnant biochemistry club, students serve as members on the department DEI committee, and we have had students become members of our HHMI cohort group. Sometimes all that is needed to light a fire is a small spark—and in our department—our community of faculty and students has ignited since our inaugural event. In this chapter, we have discussed the planning, rationale, and details of our student research celebration event (poster symposium, Engelpalooza) and a faculty toolbox of learning materials. We were able to encourage and accommodate students in belongingness in our department by reassessing and redesigning the opportunity and environment. In doing so, we have begun a snowballing effort of success of student engagement where students are better positioned to be supported throughout their academic journey and are better equipped to support one another. We have more efforts and feedback to keep improving each year and look forward to continually increasing the community and knowledge of our biochemistry department.

REFERENCES

Brownell, S.E., & Tanner. K.D. (2012). Barriers to faculty pedagogical change: Lack of training, time, incentives, and … tensions with professional identity? *CBE Life Sciences Education, 11*(4), 339-346.

Canning, E.A., Muenks, K., Green, D.J., & Murphy M.C. (2019). STEM faculty who believe ability is fixed have larger racial achievement gaps and inspire less student motivation in their classes. *Science Advances, 5*(2): eaau4734. https://doi.org/doi:10.1126/sciadv.aau4734

Casadevall, A., & Handelsman, J. (2014). The presence of female conveners correlates with a higher proportion of female speakers at scientific symposia. *MBio, 5*(1), 1-4.

Chrikov, I., Soria, K., Horgos, B., & Jones-White, D. (2020). Undergraduate and graduate students' mental health during the COVID-19 pandemic. SERU Consortium, University of California Berkeley and University of Minnesota. Retrieved April 2023, from https://conservancy.umn.edu/handle/11299/215271

Eagan, M.K., Sharkness, J., Hurtado, S., Mosqueda, C.M., & Chang, M.J. (2011). Engaging undergraduates in science research: Not just about faculty willingness. *Research in Higher Education, 52*: 151-177.

Esquivel, A., Marincean, S., & Benore, M. (2023). The effect of the COVID-19 pandemic on STEM faculty: Productivity and work-life balance. *PLoS ONE 18*(1): e0280581.

Flaherty, E.A., Urbanek, R.E., Wood, D.M., Day, C.C., D'Acunto, L.E., Quinn, V.S., & Zollner, P.A. (2018). A framework for mentoring students attending their first professional conference. *Natural Sciences Education, 47*(1): 1-8.

Gopalan, M., & Brady, S.T. (2020). College students' sense of belonging: A national perspective. *Educational Researcher, 49*(2): 134-137. https://doi.org/10.3102/0013189X19897622

Giovenco, D., Shook-Sa, B.E., Hutson, B.E., Buchanan, L., Fisher, E.B., & Pettifor, A. (2022). Social isolation and psychological distress among southern U.S. college students in the era of COVID-19. *PLoS One, 17*(12): e0279485.

Lacher, L., & Biehl, C. (2018). Using Discord to understand and moderate collaboration and teamwork. *Proceedings of the 49th ACM technical symposium on computer science education* (pp. 1107).

Merrian-Webster. (n.d.). Lollapalooza. In *Merriam-Webster.com dictionary*. Retrieved June 1, 2023, from https://www.merriam-webster.com/dictionary/lollapalooza

Mowreader, A. (2023, March 14). Campus engagement tip: Building community and engagement. *Inside higher ed.* Retrieved in April 2023 from:https://www.insidehighered.com/news/2023/03/15/six-factors-improve-student-connections

O'Meara, K.A. 2022. Enabling possibility: Reform of faculty appointments and evaluation. *TIAA Institute.* Retrieved April 10, 2023, from https://www.tiaa.org/content/dam/tiaa/institute/pdf/insights-report/2022-03/tiaa-institute-reform-of-faculty-appointments-and-evaluation-omeara-ti-march-2022.pdf

Pedersen, C.L., Lymburner, J., Ali, J.I., & Coburn, P.I. (2013). Organizing an undergraduate psychology conference: The successes and challenges of employing a student-led approach. *Psychology Learning and Teaching, 12*(1), 83-91.

Reutskaja, E., Iyengar, S., Fasolo, B., & Misuraca, R. (2020). Cognitive and affective consequences of information and choice overload. In R. Viale (Ed.), *Routledge handbook of bounded rationality* (pp. 625-636). Routledge.

Rowe, N., & Ilic., D. (2015). Rethinking poster presentations at large-scale scientific meetings – is it time for the format to evolve? *The FEBS Journal, 282*(19), 3661-3668.

Samuelson, W., & Zeckhauser, R. (1988). Status quo bias in decision making. *Journal of Risk and Uncertainty, 1,* 7-59.

Sardelis, S., & Drew, J.A. (2016). Not "pulling up the ladder": Women who organize conference symposia provide greater opportunities for women to speak at conservation conferences. *PLoS One, 11*(7): e0160015.

Soria, K., Brayden, R., Horgos, B., & Hallahan, K. (2020). Undergraduate experiences during the COVID-19 pandemic: Disparities by race and ethnicity. SERU Consortium, University of California Berkeley and University of Minnesota. Retrieved April 5, 2023, from https://conservancy.umn.edu/handle/11299/218339

Sullivan, B. (2021). When you feel overwhelmed, subtraction can lead to gains. *Psychology Today*. Retrieved April 5, 2023, from https://www.psychologytoday.com/us/blog/pleased-meet-me/202106/when-you-feel-overwhelmed-subtraction-can-lead-gains

Todorinova, L., & Wilkinson, Z.T. (2020). Incentivizing faculty for open educational resources (OER) adoption and open textbook authoring. *The Journal of Academic Librarianship, 46*(6), 102220.

Xu, D., Solanki, S., McPartlan, P. & Sato, B. (2018). EASEing students into college: The impact of multidimensional support for underprepared students. *Educational Researcher, 47*(7): 435-450. https://doi.org/10.3102/0013189X18778559

White Berheide, C., Carpenter, M.A., & Cotter, D.A. (2022). Teaching college in the time of COVID-19: Gender and race differences in faculty emotional labor. *Sex Roles, 86*(7-8), 441-455.

Zajonc, R.B. (1968). Attitudinal effects of mere exposure. *Journal of Personality and Social Psychology, 9*(2p2), 1-27.

This program is supported in part by a grant to Virginia Tech from the Howard Hughes Medical Institute through the Inclusive Excellence Grant.

VIGNETTE: REFLECTING ON THE IMPACT OF ENGELPALOOZA

ERIN DROLET

As a first-year student in biochemistry, I faced challenges connecting with other students in my major and engaging with biochemistry faculty members. Many of our introductory courses were attended by students from various disciplines, making it difficult to establish a strong sense of community within the biochemistry program during my first two years at Virginia Tech. However, in my third year, the Department of Biochemistry took steps to address this issue by organizing an annual event called Engelpalooza. This event, initially a poster session for undergraduate researchers to present their work, has evolved over the years to include guest speakers, research demonstrations, and information on numerous opportunities for undergraduate students. Personally, I presented my research at two Engelpalooza events and was involved in planning a third. Engelpalooza not only boosted my confidence in my research skills as an undergraduate student but also fostered a sense of belonging within the biochemistry community. Being part of the planning committee as a graduate student allowed me to contribute to the growth of Engelpalooza, particularly for incoming students, and further strengthen our community. This event played a crucial role in building the sense of camaraderie that was lacking when I initially joined the biochemistry program.

During the summer of 2019, I worked as an undergraduate researcher for Dr. Anne Brown and presented my initial findings at a research poster session in August. Surprisingly, I discovered a deep appreciation for attending research poster symposia. Creating a poster helped me gain a better understanding of my own findings, and I thoroughly enjoyed sharing and discussing my research with others. However, the audience at this event consisted of undergraduate students from various fields, leading to difficulties in explaining my research to those unfamiliar with biochemistry. In October 2019, I had the opportunity to present my poster at Engelpalooza, which was a vastly different experience.

Because everyone in attendance was knowledgeable about biochemistry, I could delve into more detail about my research and engage in thought-provoking discussions. Engelpalooza also provided me with the chance to interact with faculty members I had not yet met and explore research conducted in other laboratories within our department. It helped me transcend the boundaries of my own research space and truly become part of the larger biochemistry community. Engelpalooza not only served as a platform for professional development but also allowed new undergraduate students to see the excitement of research. Students attending the event could explore innovative scientific posters and engage with the presenters, igniting their curiosity about research. In fact, it was common for students to reach out to faculty members after being inspired by their work at Engelpalooza. This event played a pivotal role in breaking down barriers and demonstrating the thrilling aspects of research to prospective students, making it less intimidating for them to approach faculty members and inquire about joining their labs.

In 2020, community was lacking in many of our lives due to the pandemic. The Department of Biochemistry decided to hold Engelpalooza virtually via Discord, an instant messaging and video calling social platform where users can interact in private chats or as part of communities called "servers". Several biochemistry students and faculty members worked together to create a Discord server to host Engelpalooza. Within the server, there were voice chat rooms where undergraduate and graduate students presented their research. Students were able to experience presenting their research when many other conferences were canceled. I was one of the many students who presented at this Engelpalooza. I had already given several research presentations up to that point, so I did not grow as much professionally from this experience as I had with my first Engelpalooza. However, Engelpalooza that year tremendously boosted my morale because I was able to interact with my friends in the department who I had not spoken to since March 2020. The event helped me feel like I was not alone and made me more excited for when we could hold Engelpalooza in person once again.

In addition to the research presentation voice chat rooms, there were other spaces in the Discord server dedicated to sharing various community resources, including information about graduate school, the American Society for Biochemistry and Molecular Biology, undergraduate research, and study abroad. There were also chats to network with students and faculty. Engelpalooza helped our department feel like a community in a time of isolation. Not only was Engelpalooza important for the first-year students in our department to learn about the opportunities that are available to them and meet their peers, but it was also helpful for our continuing students and faculty members to feel that our community was still present. One of the other important things that came out of Engelpalooza was a Discord server for Biochemistry students to chat with each other, ask for help in classes, and play video games together. Overall, Engelpalooza 2020 helped to continue building our community and helped us feel connected in a time of virtual learning.

The next Engelpalooza that I was involved in was in 2022, and it was back to being in person. I was a graduate student at the time, and my role was to help coordinate the different information and activity tables. That year at Engelpalooza, there was information about the undergraduate Biochemistry Club, Biochemistry Graduate Student Association, the American Society for Biochemistry and Molecular Biology, undergraduate research, biochemistry peer mentors, and study abroad. At one table, they held a demonstration for Oxford Nanopore sequencing to get people excited about an undergraduate research experience pilot program. In a nearby room, another group was demonstrating virtual reality and how it could be used for science communication. Other members of the planning committee expanded the event to include a Virginia Tech Biochemistry alum to speak before the poster sessions, and inviting our school mascot, the Hokie Bird, to visit. Several biochemistry classes were canceled that day to enable students to go to Engelpalooza, and some professors offered extra credit for talking to some of the students presenting their research.

Engelpalooza 2022 was a resounding success, with numerous students presenting their research and many more attending to support their peers and learn more about our department. It was a delightful experience to contribute to the event's planning and ensure the best possible experience for the students. Involvement in the planning committee also revealed to me that the impact of Engelpalooza extended beyond the students. Faculty members eagerly participated in Engelpalooza, displaying enthusiasm for the impressive research conducted by our students. A few days after the event, the planning committee convened to discuss ideas for enhancing and expanding Engelpalooza for the following year. With several new members joining our department, a multitude of fresh ideas emerged to transform Engelpalooza into more than just a modest poster session for our students but a community-wide event that unites everyone.

Engelpalooza has proven to be an invaluable experience for me and many other students in the Department of Biochemistry. It offers students their initial exposure to research, which is conducted by their friends and peers. Many faculty members also attended the event. Engelpalooza demonstrates to students that research is an exciting endeavor and diminishes the intimidation of approaching faculty members to join their labs. Personally, as a presenter, I have experienced the tremendous confidence boost that comes from sharing your work for the first time. Additionally, the process of creating a poster and articulating your findings to others compels you to possess a strong understanding of your research. It is rewarding to have a tangible representation of your hard work. As an undergraduate student, Engelpalooza enhanced my skills as a science communicator and instilled excitement in me about the work my peers were undertaking. As a graduate student serving on the planning committee, I have witnessed the significant impact Engelpalooza has on the faculty within our department. Faculty members eagerly participate and encourage student involvement in the exciting opportunities our department offers. They equally relish engaging with students and learning about their research. Faculty members pose thought-provoking questions that stimulate students to think critically about their findings and potentially identify new research directions. At its core, Engelpalooza is a research poster symposium. However, it has evolved into a pivotal event for our community, prompting us to expand it each year by incorporating additional events and opportunities to foster growth and connection among our students.

This program is supported in part by a grant to Virginia Tech from the Howard Hughes Medical Institute through the Inclusive Excellence Grant.

CHAPTER 9.

TEACHING TO MAKE MATH RESONATE

Including Social Justice in a Graduate Course for Teachers
DIANA S. CHENG AND JOHN B. GONZALEZ JR.

ABSTRACT

This chapter describes how the first author ("I") introduced social justice mathematics lessons (SJMLs) to teachers taking a graduate mathematics education course on problem solving tailored to the middle and high school levels. SJML's provide students with experiences where mathematics can be used to make the world more equal and just, and learning to teach for social justice is a process by which teachers adapt their mathematical lessons to social justice contexts. In particular, I explain how I mentored teachers first to practice solving SJMLs, then use micro-teaching with their peers to create or extend SJMLs to meet the needs of their student populations, and then to reflect upon their implementations to build a sustainable way to continue using SJMLs in their instructional practices. The results of pre- and post- surveys indicated favorable shifts in teachers' beliefs about their teaching for the purposes of empowering students to consider social justice. From the lens of a facilitator of the professional learning experience, we discuss the sustainability of this approach to helping mathematics teachers teach SJMLs.

INTRODUCTION

As a step towards creating a more equitable and just society, educators have been advocating for teachers to use social justice topics within school curriculum (Darling-Hammond et al., 2017). Mathematics lessons within social justice contexts can help students become knowledgeable about social issues, connect mathematics with students' cultural and community histories, empower students to confront and solve their real-word challenges, and help students use mathematics as a tool for social change (Berry et al., 2020). In tandem, students' desires to conduct more advanced analyses to explain situations and experiences can motivate the learning of additional mathematical topics (Gutstein, 2003).

Graduate courses providing professional learning experiences can help teachers incorporate social justice mathematics lessons (SJMLs) in their teaching (Bartell, 2013). Teaching mathematics with a social justice lens is a complex process that spans beyond the short time interval of just one graduate course.

SOCIAL JUSTICE MATHEMATICS LESSONS AS AN ELEMENT OF DIVERSITY, EQUITY, AND INCLUSION

Teaching mathematics for social justice includes the idea of not just preparing students to live in their worlds, but also to help them improve and revise the social structures they experience (Bartell, 2013). Mathematics can be used to describe and analyze issues of social justice, and to support the development of actions to enact transformative changes in the real world (NCSM & TODOS, 2016). Some examples of social justice issues for which mathematical analyses can be informative include but are not limited to diversity, structural and systemic policies that reproduce inequity, and gaps in outcomes and opportunities for minoritized groups (Cochran-Smith & Keefe, 2022).

The Learning for Justice (2022) organization developed standards to assist teachers in categorizing social justice issues. The four main categories involving social justice include: Identity, Diversity, Justice, and Action. Within the Identity standard, teachers should help their students express pride and have a positive self-esteem of their own identities without degrading others. In the Diversity standard, teachers should guide students to examine social, cultural, political, and historical facets of diversity and exchange ideas and beliefs in open-minded ways. To address the Justice standard, teachers should provide tools to students to recognize when unfairness and biases exist. The Action standard includes making informed decisions about when and how to take a stand against biases and prejudices. These standards were written with all academic subjects in mind, and various types of mathematical ideas can assist in addressing these standards.

CONTEXT FOR SOCIAL JUSTICE MATHEMATICS LESSONS WITH TEACHERS

The ten teachers in the Problem Solving for Teachers graduate course were enrolled in the Masters of Science in mathematics education degree program at Towson University, a public institution in the University System of Maryland. The teachers were licensed to teach mathematics in the state and they all worked for the same public school district in the county in which the university is located, Baltimore County. Baltimore County Public Schools paid Towson University the tuition for the courses they took, based on a direct-billing arrangement that was negotiated between the district and the university. The teachers were expected to pay back the district for graduate course tuition if they failed a course.

The teachers began their master's degree program in Fall 2019, took three graduate courses per year for four years, and graduated with their degrees in Spring 2023. The master's degree was designed as a cohort, in which the teachers took all of their twelve courses with the same group of classmates. Each of the graduate courses had a different instructor.

The seven courses that the teachers took prior to Spring 2022 ranged from focusing on mathematics content (algebra, data analysis and probability) and pedagogy (Common Core State Standards, technological tools, etc.). The course that the teachers took in the semester immediately preceding this problem solving course was Makerspace Technology in the Classroom (described in further detail in the chapter "Preparing STEM Teachers to Be Change Makers" in this book). The remaining courses in their graduate course sequence from Spring 2022 through Spring 2023 included topics such as literacy, leadership in equity, understanding and using mathematics education research, and a graduate project.

The graduate course described in this chapter is the eighth in a pre-designed sequence of twelve graduate courses for this group of teachers. This course was taught in Spring 2022 during a ten-week session in January through mid-March. The first author was the instructor for this course. The course was offered entirely online, but the first author met all of the students in-person while attending an hour of their Fall 2021 semester course which was conducted in a hybrid modality. Five of the teachers were alumni of Towson University's undergraduate teacher preparation program, and the first author had existing relationships with them through prior coursework and other academic events.

Nine of the ten enrolled teachers identified as White, and one identified as Asian. In contrast, according to demographic information published by the Maryland State Department of Education (2022), approximately 33% of the enrolled students in the school district where these teachers teach are White and 7% are Asian. Other races represented in the student body of the school district include 40% African-American and 14% Hispanic. Four of the teachers (three females and one male) were full-time mathematics teachers in high schools, and five of the graduate students (three females and two males) were full-time mathematics teachers in middle schools.

WHY SOCIAL JUSTICE RELATED DISCUSSIONS MIGHT NOT BE MORE WIDESPREAD

Discussions related to social justice are not always encouraged or welcome in classrooms, even when faculty members might believe strongly that engaging in such discussions in academic settings can better prepare their students for their future careers. For example, in the state of Florida, there is legislation that restricts the teaching of ideas about race and gender in higher education; as a result, educators have changed the content of their courses when faced with the threat of losing their jobs (ACLU, 2022).

For the purposes of this course, we are fortunate that Baltimore County Public Schools is extremely supportive of initiating social justice discussions within classrooms. Dr. John Staley, a Baltimore County Public Schools mathematics educator and leader, was one of the co-editors of the course textbook that was used (Berry et al., 2020). Dr. Staley has initiated many district-wide and national professional learning experiences to help teachers better "engage students in critical inquiry about the world and potential injustices surrounding them, pushing students to imagine and create a world with justice, fairness, and equality (Berry et al., 2020, p. 25)." In fact, Dr. Staley also taught another course in the sequence of courses for this degree program in Spring 2023, focused on creating an equity improvement plan to address an issue within teachers' schools.

MATHEMATICAL MODELING AND SOCIAL JUSTICE

Mathematical modeling is one way of problem solving that involves a process of representing real-life situations (Lesh & Doerr, 2003) and is valued in the Common Core State Standards (CCSSI, 2010). Understanding the cycle of mathematical modeling is a pre-requisite to truly applying mathematics to social justice contexts. Modeling is similar to scientific experimentation in that it begins with a key question that the student is trying to address using mathematics (Hirsch & Roth McDuffie,

2016). The question could be an open-ended question for which the instructor does not know the answer, or one that has a definite solution that other researchers have already found. But in either case, the student engaging in the mathematical modeling should not already know the answer to the question.

The student actions that take place while solving a mathematical modeling problem related to a social justice context, were described by Jung & Brand (2021). At the outstart, students need to interpret the key question and the underlying social justice issue. Sometimes, in order to understand the problem, it may be illuminating to review multiple modes of media such as videos, news articles, or tables of data related to the context. Next, students would propose some approaches to addressing the social justice problem. This may include researching additional information that could help solve the problem. Next, students would need to mathematize the situation, such as writing mathematical equations or developing other representations of the context. Next, students would validate the solution against contextual constraints (e.g., considering whether the solution makes sense or is acceptable for the given situation). Finally, student would apply the solution to the social justice context. This may include justifying mathematical conclusions as they relate to the context, and reporting the conclusions so that people can affect informed actions.

Mathematical modeling is designed to be an iterative process (Anhalt & Cortez, 2015). Once a student undergoes the process once, they may re-evaluate whether the solution is the best fit for the situation and consider taking into account additional constraints in the next iteration of the cycle to arrive at a conclusion that is more realistic or viable.

WHY SOCIAL JUSTICE MATHEMATICS LESSONS MATTER

Berry et al. (2020) provided several reasons as to why mathematics educators should use social justice contexts in classroom lessons. First and foremost, SJMLs can help build a society in which students are informed about their own lives as well as the lives of others. Mathematical analyses can help students become aware of injustices that are present and determine that they hold misconceptions about important issues. Second, SJMLs can connect with students' cultural and community contexts, and thus provide motivation and appreciation for learning mathematical content. Third, SJMLs can help empower students to identify issues and propose solutions to real-world challenges that they face. Fourth, SJMLs help students learn first-hand that mathematics can be a valuable tool used to inform social change.

IMPLEMENTATION OF THE GRADUATE COURSE

There were two main textbooks for the course, edited by Gutstein & Peterson (2013) and Berry et al. (2020). Both of these texts were written by in-service teachers and provide many examples of SJMLs used in their classrooms. The texts also describe students' discussions on the issues, students' mathematical work, teachers' analyses of the student work, and student feedback on the lessons. The texts include authors' recommendations for future implementations of their lessons. Berry et al.'s (2020) book includes a link to online resources, such as presentation and spreadsheet files that teachers can easily use to conduct the lessons in their classrooms. Berry et al.'s (2020) book also provides a lesson planner template that teachers can use to plan social justice lessons on other topics, and recommendations for teachers for conducting discussions on controversial social justice topics. We found these texts to be useful for our graduate students, as they conveyed both enthusiasm and the importance of raising social justice issues with grade school students and showed how mathematics was used in the lessons.

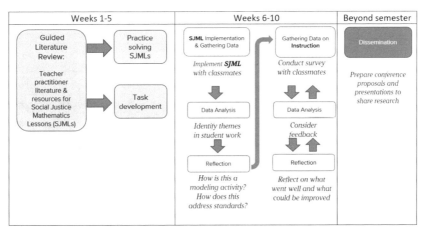

Figure 9.1: Key features of Problem Solving for Teachers Course

Figure 9.1 shows a visual diagram with the key features of this course, intended to be a mentoring cycle for teachers—a modified version of the mentoring cycle developed by (Kara & Corum, 2023)—to become more familiar with conducting (SJMLs). The first half of the course included a guided literature review, where graduate students learned about the Social Justice Standards (Learning for Justice, 2022) and, culturally relevant mathematics pedagogy (Aguirre & Del Rosario Zavala, 2013) because this includes the use of social justice contexts in mathematics lessons, and solved SJMLs that were designed for middle and high school students. The majority of the activities in which graduate students participated during the in-class sessions were adapted from Gutstein & Peterson (2013), and a few were adapted from Berry et al. (2020). One lesson on diversity in cultural contexts was adapted from Goffney & Gutierrez (2018). Appendix A describes how a geometry lesson was adapted for online instruction.

The first author also used an originally designed SJML, which is further described in Cheng (2023a). Because the 2022 Winter Olympic Games took place during the semester in which this course was taught, the first author used the mathematical tools of power indices to describe an unjust situation in a current event. In the figure skating team event, the Chinese team included an American-born female skater Zhu Yi who renounced her United States citizenship in order to represent China. Yi earned last place in the first round of the team event, and Chinese citizens criticized her heavily on social media for the poor performance. The teachers were asked to employ two measures of power to explain that Yi's contributions in the first round did actually contribute towards the Chinese team's advancing to the second round of the team event.

As asynchronous work outside of class time, the teachers were assigned to read from both course texts and reflect on the use of SJMLs in the classroom, through discussion board posts and responses to classmates' comments. The teachers were also asked to analyze the lesson plans corresponding to the lessons they completed during the synchronous sessions using the Culturally Relevant Mathematics Teaching tool (Aguirre & Del Rosario Zavala, 2013). Additionally, the teachers completed SJMLs created by former graduate students who pre-recorded their lessons.

In the second half of the course, the teachers worked in small groups to design and implement an SJML. They were allowed to adapt a lesson from an existing source, but they were also asked to make the lesson relevant for their students by using their school data and more updated information than might have been available to the textbook authors. For example, to adapt a lesson on the Pythagorean theorem as it relates to wheelchair ramps (Gutstein & Peterson, 2013, pp. 136-137), the teachers took photographs of the various ramps at their schools and local businesses so that their students could evaluate whether these ramps meet accessibility recommendations.

Teachers also provided input to their peers on how they might revise their lessons through a feedback survey. The presenting teachers were asked to analyze their classmates' work, their classmates' survey results, and reflect on their implementation. The complete description of the final project is in Appendix B.

WHAT WE LEARNED FROM THIS COURSE

The Learning to Teach for Social Justice – Beliefs survey (Enterline, et al., 2008; Ludlow et al., 2008), abbreviated here as LTSJ-B, was initially developed to gauge pre-service teachers' beliefs when they entered and exited their four-year teacher preparation programs, as well as after their first year of teaching. When Enterline et al. (2008) and Ludlow et al. (2008) were validating their surveys, they did not use the surveys with the same group of students in a pre- and post- setting, rather they used three different groups of students in the different phases of their educational sequence of courses. In our setting, the graduate students were surveyed at two points in time: prior to the start of the course in January 2022, and in the tenth week of the course in mid-March 2022. All ten teachers enrolled in our course took the pre-survey, but only nine of the teachers completed the post-survey; thus only these nine teachers' results are reported in this manuscript.

The LTSJ-B survey consists of twelve Likert-style statements regarding beliefs about teaching, which participants were asked to rate on a scale of one to five, with the following categories: "1 = Strongly Disagree, 2 = Disagree, 3 = Uncertain, 4 = Agree, 5 = Strongly Agree." The higher the score, the more the survey participants agreed with the statements. The minimum score was twelve points, and the maximum score on the survey was sixty points. Figure 9.2 shows the pre- and post-average scores from the participants in this study. The pre-test average score was 45.56 points, indicating that the teachers already adhered to many of the desired beliefs related to learning to teach for social justice prior to the start of the course. The post-test average score was 47.11 points, which shows that there was an overall increase in teachers' beliefs in a relatively short period of time.

Five of the LTSJ-B statements were positively phrased and Enterline et al. (2008) found that pre-service teachers in their sample more readily agreed with these items. An example of one of these statements is "An important part of learning to be a teacher is examining one's own attitudes and beliefs about race, class, gender, disabilities, and sexual orientation (Enterline et al., 2008, pg. 275)." Based on their experimental data, the authors called this group of items a "non-controversial" set of items (abbreviated as "NC" in Figures 9.2, 9.3, 9.4, and 9.5), with a maximum score of twenty-five points. Seven of the items were slightly more difficult for pre-service teachers to endorse, so the authors considered these more "controversial" items (abbreviated as "C" in Figures 9.2, 9.3, 9.4, and 9.5). These "controversial" items were negatively phrased and Enterline et al. (2008) suggested that they required reverse scoring. An example of such a statement is "Economically disadvantaged students have more to gain in schools because they bring less into the classroom (Enterline et al., 2008, pg. 275)." Participants who strongly disagreed with all of the "Controversial" subset of statements would have a LTSJ-B maximum score of thirty-five points.

Figure 9.2 shows the pre- and posttest- scores on both the Non-controversial and the Controversial items from the teachers in this study. Teachers exhibited a slight increase in the average subscores on Non-controversial items (2.11 points higher on the post-test subscore), but showed a slight decrease in the subscores on the Controversial items of 0.56 points.

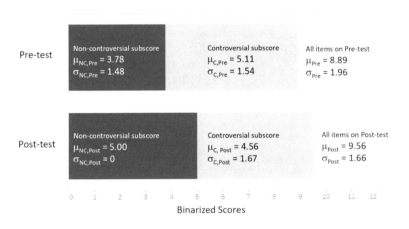

Figure 9.2: Learning to Teach for Social Justice – Belief Scores of Participants

In order to further illustrate the differences between teachers' pre- and posttest- scores, we created Figure 9.3 to show each individual student's scores. The majority of the teachers (listed as #2, #4, #5, #7, #9) had post-test scores that were higher than the pre-test. One teacher (#3) had the same score on both the pre- and post-tests, and three teachers (#1, #6, #8) had lower post-test scores.

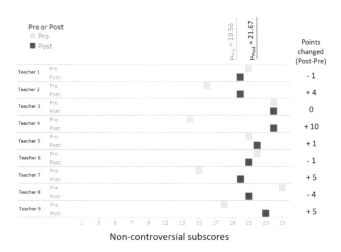

Figure 9.3: Individual scores on LTSJ-B Non-controversial subscores, pre- and post-test

RESULTS FROM BINARY SCORING OF LTSJ-B SURVEYS

We also re-scored the teachers' Likert ratings on the LTSJ-B survey in a binary method, because we wanted a way to detect whether teachers shifted away from disagreeing with the items between the pre- and post-test administrations. In the original rating scale, if teachers rated an item as either 1 = Strongly Disagree, or 2 = Disagree, we assigned a score of zero to this item. All other ratings (3 = Uncertain, 4 = Agree, 5 = Strongly Agree) were assigned a score of one. Using this binary scale, the minimum score was zero if teachers disagreed with all twelve statements, and the maximum score was twelve if teachers didn't disagree with all twelve statements.

Figure 9.4 shows results of this binary rescoring. We observed an increase in the overall average scores, from the pre-test average score of 8.89 points to the post-test average score of 9.56 points. For the non-controversial (NC) statements, the average subscore on the pre-test was 3.78 points, and the average subscore on the post-test was 5 points. In other words, all of the teachers agreed with the NC items after completing this graduate course. The average Controversial subscore decreased by 0.55 points.

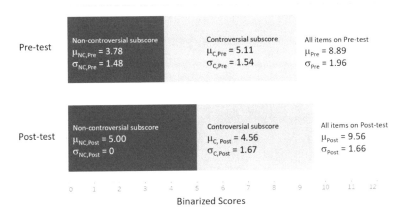

Figure 9.4: Learning to Teach for Social Justice – Beliefs binarized scores of participants

On the Non-controversial items, the majority of the teachers (#1, #2, #4, #7, #9) had a pre-test score of four or lower, but all teachers had a post-test score of five points. The alignment of all teachers' Non-controversial post-test scores is shown clearly in Figure 9.5; this is also reflected in the standard deviation of zero.

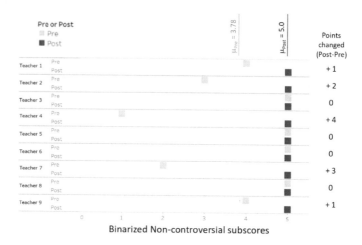

		Pre or Post	$\mu_{Pre} = 3.78$	$\mu_{Post} = 5.0$	Points changed (Post-Pre)

Figure 9.5: Individual binary scores on LTSJ-B Non-controversial subscores, pre- and post-test

STUDENT PERSPECTIVES

DISCUSSION ON LTSJ-B SURVEY RESULTS

The authors of the LTSJ-B survey strongly recommend that teacher education programs that include social justice agendas also use multiple ways to measure their students' outcomes, and not solely rely on their beliefs survey (Cochran-Smith et al., 2010). We also examined university-distributed course evaluation responses of the graduate students. The increased overall LTSJ-B survey results are consistent with one teacher's comment on a course evaluation question asking about what they liked about the course: "One of the most valuable things about this course was the chance to talk about our experiences and hopes in melding social justice into the math content area. It's not an easy thing to make teachers feel comfortable talking about those concepts, but this course was an excellent experience." In response to a course evaluation question about whether they would or would not recommend the course to others, one teacher mentioned, "Every teacher should have training to be able to speak with their students about social justice issues."

There may be several explanations for the lowered LTSJ-B Controversial item subscores on the post-test. On the course evaluation, one of the teachers wrote in response to a question about ways to improve the course: "One of the only things I could think of would be more 'baby step' lessons. For teachers not used to teaching social justice concepts, it might be easy to explore less 'hot' like race and socioeconomic issues at first and ease into the idea with things like environmental action." The teacher directly mentioned that having a hierarchy of topics that range from easy to discuss to more challenging issues would be helpful, which is consistent with having scores increase on the Non-controversial items yet not on the Controversial items.

Another explanation related to the elevated post-test Controversial item subscores might be reliability of teachers' interpretations of the statements. We found that the wording of some of the reverse-scored items was somewhat open to interpretation, so we were not too concerned about the slight decrease. For example, for the item, "Whether students succeed in school depends primarily on how hard they work" (Enterline et al., 2008, p. 275), we could interpret it such that teachers would want to have their students work hard but also see that the opportunities (or other factors such as resources, teachers' hard work, etc.) presented to their students are also a large factor of their success, and this would be consistent with Enterline et al.'s (2008) rating of a reverse scored item. However, we could also agree with a similar idea to what is stated, which is that if students aren't working very hard in school, then we might not expect for students to have successful scholastic results independently of other factors – and then this line of reasoning would support positively scoring this item. The ambiguity of the interpretation of several statements could lead someone to respond one way in one setting and an opposite way in a different point in time.

DISSEMINATION OF SJMLS BEYOND SCOPE OF GRADUATE COURSE

Two of the groups were able to disseminate their final project SJML beyond the scope of the course. One group of high school teachers compared the demographics of the students in their schools enrolled in advanced track courses with demographics of students enrolled in standard track courses, using the ideas from Berry et al. (2020, p. 81-89). One of these teachers was asked by the department chair at his school to present his findings at a faculty department meeting, and this indicates that there is strong support from the local school district for having faculty explore these issues and share their results.

Another group of teachers, Mark Heath, W. Brooks Lynch, and Jodie Wohlfort, used the ideas behind a lesson (Berry et al., 2020, p. 89-98) quantifying the number of students in a given school who might identify as lesbian, gay, bisexual, transexual, and queer (LBGTQ+). The lesson was initially written to include matrix multiplication, but the concept of matrix multiplication was too advanced for the teachers' middle school students, so the teachers adapted this lesson using multiplication of percentages so that they would be able to implement it with their students. Students were asked to research the student population in each of the grades at their school. They were then asked to apply national percentages of students who might have felt bullied or harassed due to their identities, so that they could generate an estimate of how many students attending their school might have these experiences. This group presented their work with others from Towson University at the regional conference of the National Council of Teachers of Mathematics (Ciscell et al., 2022).

As Bartell (2013) mentioned, it is impossible to address all aspects of social justice in one single graduate course. We are fortunate that several of the graduate faculty at this university are teaching with the lens of social justice, and this allows the teachers in our masters degree program to gain exposure from multiple perspectives on how to use social justice in their classroom instruction. As teachers continue to take our courses, we can use previous teachers' work to show the next group of teachers how SJMLs were used locally by their peers.

After teaching this course in Spring 2022, my colleagues and I have implemented a few other ways to broaden the audience of teachers who are exposed to SJMLs. In the Fall 2022 semester, we partnered with other Towson University mathematics education faculty and Towson University's Center for STEM Excellence to offer a fifteen-hour, five-week long online professional learning experience. This experience was approved by the Maryland State Department of Education for participants to earn one Continuing Professional Development credit toward the renewal of their Maryland Educator Certificate (statewide teacher licensure). This professional learning experience allowed teachers across the state, all of whom were not currently enrolled graduate students, to hear about how SJMLs can be used in grade school curricula. Several graduate students from Spring 2022 shared their SJML final projects during synchronous sessions of this professional learning experience, which allowed them to see how other in-service teachers interacted with the SJMLs they developed.

Another venue for expanding the reach of SJMLs is through pre-service teacher education. I have uploaded teachers' SJMLs from this course and previous semesters' courses in an online shared repository of resources for colleagues teaching undergraduate students preparing to become teachers at our university – some of these activities are also available through (Cheng, 2023b). These SJMLs have been used as asynchronous assignments, to further impact the future generation of teachers.

DISCUSSION AND FUTURE RESEARCH

In order to expand the teaching of mathematics for social justice in schools, Leonard et al.'s (2010) have suggested that mathematics teacher educators should provide and find more appropriate examples of SJMLs. Also, they suggest that mathematics teacher educators provide teachers with ways to see SJMLs in practice with the mathematical content they need to teach, and provide teachers with chances to learn from their practice to increase teachers' self-efficacy for teaching SJMLs.

Whenever possible, having community-based or school-based field experiences where pre-service and in-service teachers can field-test their SJMLs can also be illuminating. Teacher interns in Leonard et al. (2010)'s study found that providing classroom discussions and reflections after implementing SJMLs helped students realize their own agency and find their voice in working towards social change.

One semester may be too short of a time span to see significant changes in teachers' reactions to novel approaches to teaching. For example, Al Salami et al (2017) designed a fifteen-week professional learning experience on an interdisciplinary science, technology, engineering and mathematics (STEM) problem unit with middle and high school teachers, and did not find measurable changes in teachers' attitudes within the time span of the course (e.g., Al Salami et al., 2017). In the present study, while we did find measurable changes in teachers' responses to the non-controversial items on the LTSJ-B survey, teachers did not increase in their beliefs on the more controversial items within the time span of this course.

Future research could also involve a more comprehensive evaluation of teachers' learning to teach for social justice. Guskey (2002) suggested several general levels of professional development evaluation that can apply to any kind of intervention aimed to help teachers improve their teaching. The first level of evaluation includes measuring participants' reactions to the professional development through questionnaires to gauge their initial satisfaction. The second level of evaluation involves measuring knowledge and skills that participants gained through a content assessment. The third level of evaluation includes examining organizational policies and practices that may assist or undermine implementation efforts (in our social justice example, this could include whether the school principal or district supports this type of instruction). The fourth level of evaluation investigates whether the participants' new knowledge or skills have made a difference in their professional practices – and enough time must pass to allow participants to include this new knowledge into their teaching. The fifth and final level of evaluation examines whether student learning outcomes have changed as a result of teachers' novel instructional approaches. Developing a longer term and more comprehensive support system for teachers to increase their capacity to teach social justice within mathematics classes is one of our professional goals.

As of July 1, 2023, the first author is one of the co-Principal Investigators on a project entitled "Making STEM Matter: Transforming Learning through Teacher Leadership, Justice-Centered Pedagogy, and Makerspace Technology," which will be supported by the National Science Foundation's Robert Noyce Teacher Scholarship program (Award #2243461). This project aspires to support STEM teacher leaders to develop equity-focused and social justice-oriented pedagogy. This new project also aspires to create a repository of maker-enhanced STEM lessons that can be shared more widely with STEM teachers beyond the partnering school district. The time frame of this project is five years, which includes teacher leaders' participation in several different professional learning experiences from different perspectives – both as graduate students taking graduate courses, and as instructors of courses which other teachers are taking for Continuing Professional Development.

Acknowledgement

This project is based upon work supported by the National Science Foundation under Grant No. 2243461. Any opinions, findings, and conclusions or recommendations expressed in this material are those of the author(s) and do not necessarily reflect the view of the National Science Foundation.

APPENDIX A

EXAMPLE OF SOCIAL JUSTICE MATHEMATICS LESSON USED IN THE GRADUATE COURSE, PROBLEM SOLVING FOR TEACHERS

The SJML described in this appendix was implemented with the Problem Solving for Teachers graduate students in Spring 2022. According to the Learning to Teach for Justice (2022) standards, students should be able to recognize or describe unfairness in many forms (Justice 12 / JU.6-8.12). Mathematics can be a useful tool to help students explore whether geographic representations that they see are fair or accurate.

CONTEXT

Two of my former graduate students, who were sixth grade teachers, adapted the lesson "Math, Maps, and Misrepresentation" lesson (Gutstein & Peterson, 2013, p. 189-194) for their final project in a Spring 2020 section of Problem Solving for Teachers. In Fall 2021, two of my undergraduate pre-service teachers used those lesson plans, further revised the lesson, and implemented it with middle school students. These middle school students voluntarily attended a free enrichment activity at our university's campus, and the student work from this implementation was shown to the graduate students after they participated in problem solving of this task. All groups of activity participants compared the Mercator map projection of the earth with an internet-based map projection and used proportional reasoning to better understand relative sizes of countries. I compiled all of the versions of this lesson, including middle school student work samples, to show to my Spring 2022 Problem Solving for Teachers graduate students.

TASK DESCRIPTION

Mathematics is inherent in making maps. All maps are misleading because they are two-dimensional representations of a three-dimensional Earth. In this activity, students will examine three map projections: the 1569 Mercator projection, the 1974 Peters projection, and the True Size (n.d.) projection. Students will compare the relative sizes of countries in these map projections. Mathematical content involved in this activity include finding areas of irregular shapes (HSG.MG.A.1 in the Common Core State Standards, CCSSI, 2010), using proportional reasoning (7.RP.A.2 in the Common Core State Standards, CCSSI, 2010), and interpreting scaled drawings (6.RP.A.3 in the Common Core State Standards, CCSSI, 2010).

INSTRUCTIONAL GOALS.

STUDENTS WILL...

Construct two-dimensional representations of a sphere to understand why land masses near the Earth's poles are more distorted than land masses near the Equator.

Determine areas of land masses taken from various map projections

Discuss why using misleading maps (e.g., having certain land masses under-represented or over-represented in area) in schools is a social justice issue

Student Outcomes. Students were able to distinguish between Mercator, Peters, and True Size (n.d.) projections. Using counting strategies, students were able to determine the areas of irregular figures. Students analyzed differences in area of two land masses (Greenland and Africa) to provide a mathematical rationale for why the Mercator and Peters projections are problematic.

Figure 9.6: Photographs of student work on the Math, Maps, and Misrepresentation activities

Activity 1: Covering a Styrofoam ball representing the earth	Activity 2: Flattened out paper that covered the Styrofoam ball representing the Earth

Note: Lines of longitude and latitude are drawn on the paper that is wrapped around the Styrofoam ball, with the goal of comparing these lines when the paper is flattened.	Note: The lines of longitude / latitude around the equator remain intact on this unfolding of the paper, whereas the lines near the North and South poles are more spread out when the paper is flattened
Activity 3: Student Work on Mercator Projection showing that Greenland's area seems larger than Africa's area	Activity 4: True Size (n.d.) Projection showing that Greenland's area is much smaller than Africa's area
Student used a coordinate grid overlaying a Mercator projection of Greenland and Africa (Gutstein & Peterson, 2013, p. 192). There were options to use dynamic geometry tools or transparencies with coordinate grids on them. Greenland 141 units, Africa 130 square units 141/130 ~ 1.08 Interpretation: On the Mercator projection, Greenland appears to be slightly larger than Africa.	Students' ratio comparing the size of Greenland to the size of Africa: 836,330 sq miles to 11,670,000 square miles 836,330/11,670,000 ~ 0.072 Interpretation: On the True Size projection, Africa is approximately 14 times larger than Greenland.

Social Justice Connection

Prior to this class session, most of the graduate students did not consider the question about why cartographers might have a hard time creating a two-dimensional representation of the world. Many of the graduate students remembered having pull-down Mercator projection maps of the world in their grade schools when they were younger, but did not remember having discussions with their teachers about how these maps might be inaccurate. The students seemed surprised that the misrepresentations on the map were so egregiously departed from reality (for example, the Mercator projection's ratio is fifteen times larger than the True Size (n.d.)'s projection ratio representing the sizes of Greenland and Africa).

The graduate students began examining whether other pairs of countries would also be distorted in their relative sizes, to see whether the same effect would hold for other regions of the world. One of the graduate students completed an undergraduate study abroad trip to Peru, so she was interested to see how other countries compared to Peru. She noted, "Peru & Sweden look about the same on the Mercator projection, but when you drag one over the other, Sweden is actually much smaller." Other students noticed that "Canada looks much larger than Africa, but it is actually smaller."

Students also commented on the perspective of the Mercator projection, in that the center of the map was in Europe (and they mentioned that they have also seen other projections of maps with other countries in the center, based on the country of origin of the creators of the maps). Perhaps there is nothing inherently incorrect about centering the map on one's own country of origin, but being aware of this bias is important to note. Any map can be recreated to have a different country at its center.

There was also some discussion about how far the countries are from the Equator or the North or South Poles, and the amount of distortion that might take place. Based on students' paper-wrapping activity, they noticed that any country that is closer to the poles might easily get stretched to appear much larger than reality on a two-dimensional projection. Being aware of this challenge is also illuminating.

This lesson shows how students can use ratios to explain the relative sizes of land masses in two-dimensional projections of the world. This, in turn, helps quantify how the maps are biased. The activity relates to the Justice standard #13, "Students will analyze the harmful impact of bias and injustice on the world, historically and today." For example, using a distorted map can negatively impact resource allocation decisions if people underestimate the size of the African continent.

APPENDIX B

FINAL PRESENTATION – SOCIAL JUSTICE MATH LESSON DESCRIPTION

Note: The grade on the final presentation (between the implementation and reflections written afterwards) comprised approximately 20% of the overall course grade.

You will develop a social justice math lesson to present individually or with a small group. You may use a lesson idea from a published resource (e.g., a book chapter or an article), and adapt it to a local context or a context that is of greater interest to your students. If you choose a lesson idea from a published resource, please find a context / topic that is different from a chapter covered within this class.

Please plan to implement this as an in-class synchronous activity during your assigned class session. Each presentation will span approximately 75 minutes, including time to have your classmates complete the activity. You should select 1-2 of the activity resources that are available from the book chapter to present to the class. Even distribution of workload among group members will be considered in the grading of the presentation. Non-presenters (your classmates who are participating in your activity) will need to submit their completed version of your handout.

PRODUCT 1. SOCIAL JUSTICE MATH LESSON PRESENTATION: BOOK CHAPTER AND INSTRUCTIONAL ACTIVITY

Address the following questions (not necessarily in this order) in your presentation:

- What are the recommendations of the authors of the chapter, especially related to social justice / equity / cultural identity and mathematics learning?

- Select one activity that the authors described and have your classmates participate in the activity (you will need to adapt the activity for virtual participation). Classmates should solve math problems that the students in the chapter were asked to solve, and possibly also solve extension problems that you or the authors create to further illustrate the ideas behind the chapter.

- Explain how the authors recommend that teachers implement the selected activity with grade school students.

- Find at least one additional resource beyond the book chapter that could help you answer the following questions within your presentation:

- How can you introduce the context of the activity to the students? Perhaps you can use an engaging video or children's book or other resource.

- How can you scaffold the activity for diverse learners? This can be a combination of your own ideas and the authors' recommendations, as well as via resources that you find.

- What are some anticipated student challenges with the problems? [either explicitly stated in the chapter itself, or some that you can anticipate based on your experiences in the classroom]

- A summary of Common Core State Standards – Content Standards (CCSSI, 2010) that could be addressed during the solving of the activities presented

- Analyze the strengths and weaknesses / room for improvement of the authors' lesson based on the Culturally Relevant Mathematics Teaching Lesson Analysis Tool (Aguirre & Del Rosario Zavala, 2013).

PRODUCT 2. CLASSMATES' PARTICIPATION HANDOUT

Your handout (or GoogleSlide Deck) should provide space for your classmates to take notes during your scheduled presentation (it could also list problems to be solved, or be a worksheet taken directly from the online resource). The handout should act as a reference page for your classmates.

HANDOUT PART 1: STUDENT HANDOUT TO BE ASSIGNED TO CLASSMATES AS PARTICIPATION WORK

An open-ended, cognitively challenging problem or set of problems (that could have multiple correct solutions depending on choices of constraints) that extends the original modeling problem, should be provided to students to complete individually. The problem could involve internet research or other research beyond the scope of what is being presented in class. You will be collecting your classmates' work on these problems and analyzing the work for Product 4.

HANDOUT PART 2: ANSWER KEY

The presenting group should submit a set of sample complete solutions of the problems (at least two mathematically different approaches / data examined, or two versions of constraints), along with an analysis of strengths & weaknesses of the model you developed.

PRODUCT 3: INDIVIDUAL REFLECTION ON PRESENTATION(15 POINTS)

Following the presentation, each presenter will write an individual reflection covering the following topics:

- Informed by your classmates' feedback on your lesson, discuss the parts of the lesson that went well and how to improve the parts with which you are not satisfied.

- Describe the part of your lesson that you feel was most successful and the part that needs the most improvement. You may use the Culturally Relevant Mathematics Teaching Lesson Analysis Tool (Aguirre & del Rosario Zavala, 2013) to assist you in your analysis.

What would you change for future implementations of this activity?

What would you change if you were to implement this with grade school students?

What did you learn in the process of preparing and giving the presentation?

Please rate the distribution of workload among all group members. You should provide qualitative descriptions of what each group member contributed, as well as give a percentage estimate of the workload completed by each group member so that the total workload completed adds up to 100%.

PRODUCT 4: GROUP REFLECTION (15 POINTS):

Refer to the National Council of Teachers of Mathematics book (Hirsch & Roth McDuffie, 2016). Include an explanation of how your activity fits the criteria listed in these chapters:

- Chapter 16, about what Modeling Tasks look like
- Chapter 8, about features of Modeling tasks that support students engaging with the real world in authentic ways

PRODUCT 5: CLASSMATES' WORK ANALYSIS (15 POINTS)

As a group, examine your classmates' solutions to the activities/ problems that you wrote. Develop a rubric to score your classmates' solutions – the participation work should be scored out of 15 points.

- For any incorrect solutions, determine what the error(s) were and classify the solutions based on error type.
- Classify all of the correct & incorrect solutions by type of solution method used.
- Prepare 1-3 slides (add to your existing presentation slides) clearly explaining the different types of correct solutions and incorrect solutions. [These slides could be used in future professional development sessions that you may give related to this activity.]

REFERENCES

Al Salami, M. K., Makela, C. J., & De Miranda, M. A. (2017). Assessing changes in teachers' attitudes toward interdisciplinary STEM teaching. *International Journal of Technology and Design Education, 27*, 63-88.

American Civil Liberties Union. (2022, August 17). Florida educators and students file lawsuit challenging "Stop W.O.K.E." censorship law. https://www.aclu.org/press-releases/florida-educators-and-students-file-lawsuit-challenging-stop-woke-censorship-law

Anhalt, C. O., & Cortez, R. (2015). Mathematical modeling: A structured process. The Mathematics Teacher, 108(6), 446–452. https://doi.org/10.5951/mathteacher.108.6.0446

Aguirre, J. & Del Rosario Zavala, M. (2013). Making culturally responsive mathematics teaching explicit: A lesson analysis tool. *Pedagogies: An International Journal, 8*(2), 163-190. http://dx.doi.org/10.1080/1554480X.2013.768518.

Bartell, T. G. (2013). Learning to teach mathematics for social justice: Negotiating social justice and mathematical goals. *Journal for Research in Mathematics Education, 44*(1), 129–163. https://doi.org/10.5951/jresematheduc.44.1.0129

Berry III, R., Conway IV, B., Lawler, B. & Staley, J. (2020). *High school mathematics lessons to explore, understand, and respond to social injustice*. Corwin Press.

Cheng, D. (2023a). Using the Olympics to teach mathematical modeling and social justice. *Banneker Banner, 35*(1), 3-10.

Cheng, D. (2023b). MATH 325/525: Problem solving for middle school teachers. CUREnet: Course-based Undergraduate Research Experience. Retrieved May 27, 2023, from: https://serc.carleton.edu/dev/curenet/collection/264808.html

Cochran-Smith, M., & Keefe, E. S. (2022). Strong equity: Repositioning teacher education for social change. *Teachers College Record, 124*(3), 9–41. https://doi.org/10.1177/01614681221087304

Cochran-Smith, M., Gleeson, A. M., & Mitchell, K. (2010). Teacher education for social justice: What's pupil learning got to do with it? *Berkeley Review of Education, 1.* https://doi.org/10.5070/b81110022

Common Core State Standards Initiative (CCSSI). (2010). *Common core state standards for mathematics.* Washington, DC: National Governors Association Center for Best Practices and the Council of Chief State School Officers. http://www.thecorestandards.org/wp-content/uploads/Math_Standards1.pdf

Corum, K. & Nichols, L. (2023). Preparing STEM Teachers to Be Change Makers. In J. Briganti, J. Sible, & A. M. Brown (Eds.), *Fostering communities of transformation in STEM higher education: A multi-institutional collection of DEI initiatives* (pp.13-25). Virginia Tech Publishing.

Darling-Hammond, L., Hyler, M. E., Gardner, M. (2017). *Effective teacher professional development.* Learning policy institute. https://learningpolicyinstitute.org/sites/default/files/product-files/Effective_Teacher_Professional_Development_REPORT.pdf

Enterline, S., Cochran-Smith, M., Ludlow, L. H., & Mitescu, E. (2008). Learning to teach for social justice: Measuring change in the beliefs of teacher candidates. *The New Educator, 4*(4), 267–290. https://doi.org/10.1080/15476880802430361

Goffney, I., Gutierrez, R., Boston, M. (Eds.). (2018). *Annual Perspectives in Mathematics Education: Rehumanizing mathematics for Black, Indigenous, and Latinx students.* National Council of Teachers of Mathematics.

Guskey, T. (2002). Does it make a difference? Evaluating professional development. *Educational Leadership, 59*(6), 45-51.

Gutstein, E. (2003). Teaching and learning mathematics for social justice in an urban, Latino School. *Journal for Research in Mathematics Education, 34*(1), 37-73. https://doi.org/10.2307/30034699

Gutstein, E. & Peterson, B. (2013). *Rethinking Mathematics: Teaching Social Justice by the Numbers* (2nd ed). Rethinking Schools Publication. https://rethinkingschools.org/books/rethinking-mathematics-second-edition/

Heath, M., Ciscell, J., Corum, K., Lynch, W., & Wohlfort, J. (2022, November 30). Engaging with social justice mathematics modeling lessons [Conference session]. National Council of Teachers of Mathematics Regional Conference, Baltimore, MD, United States. https://www.nctm.org/uploadedFiles/Conferences_and_Professional_Development/Regional_Conferences_and_Expositions/Baltimore2022/Schedule/Baltimore2022_Program.pdf

Hirsch, C. & Roth McDuffie, A. (Eds.). (2016). *Annual Perspectives in Mathematics Education: Mathematical modeling and modeling mathematics*. National Council of Teachers of Mathematics.

Jung, H., & Brand, S. (2021). Student actions for social justice-oriented mathematical tasks. *Mathematics Teacher: Learning and Teaching PK-12, 114*(5), 378–385. https://doi.org/10.5951/mtlt.2020.0133

Kara, M., & Corum, K. (2023). Pre-service teachers as researchers: A mentorship model. *International Journal of Education in Mathematics, Science, and Technology (IJEMST), 11*(1), 237-251. https://doi.org/10.46328/ijemst.2351

Learning for Justice. (2022). Social justice standards: The Learning for Justice Anti-Bias Framework, Second Edition. The Southern Poverty Law Center. https://www.learningforjustice.org/frameworks/social-justice-standards

Lesh, R. & Doerr, H. M. (Eds.). (2003). *Beyond constructivism: Models and modeling perspectives on mathematics problem solving, learning, and teaching*. Lawrence Erlbaum Associates Publishers.

Leonard, J., Brooks, W., Barnes-Johnson, J. & Berry, R. Q. III. (2010). The nuances and complexities of teaching mathematics for cultural relevance and social justice. *Journal of Teacher Education, 61* (3), 261-270.

Ludlow, L. H., Enterline, S. E., & Cochran-Smith, M. (2008). Learning to teach for social justice-beliefs scale: An application of Rasch Measurement Principles. *Measurement and Evaluation in Counseling and Development,* 40(4), 194–214. https://doi.org/10.1080/07481756.2008.11909815

Maryland State Department of Education. (2022). *Welcome to the Maryland report card.* Maryland State Department of Education. https://reportcard.msde.maryland.gov/

National Council of Supervisors of Mathematics (NCSM) and TODOS: Mathematics for ALL. (2016). Joint position statement: Mathematics education through the lens of social justice. Retrieved March 23, 2023, from https://www.todos-math.org/socialjustice

The True Size. (n.d.). *The true size of...*Retrieved March 23, 2023, from http://www.thetruesize.com

---

This program is supported in part by a grant to Virginia Tech from the Howard Hughes Medical Institute through the Inclusive Excellence Grant.

CHAPTER 10.

STRATEGIES FOR CREATING AND SUSTAINING INCLUSIVE MAKERSPACES

LYNN NICHOLS AND KIMBERLY CORUM

ABSTRACT

Although the world is amidst a technological renaissance, inequity prevails throughout the design, implementation, and function of modern technologies. Though not a panacea, a greater representation of Black, Indigenous, and People of Color (BIPOC) scientists, engineers, mathematicians, and computer programmers in the science, engineering, technology, and mathematics (STEM) workforce will help safeguard against digital racism. Due to the prolific nature of makerspaces in schools, one approach to combat this gap is to increase the participation of BIPOC students in school makerspaces, which provide training on a variety of STEM tools and technologies and may increase BIPOC students self-identifying with STEM professions. The purpose of this work is to explore practices that increase BIPOC student positive perception and comfort in makerspaces so that students feel empowered to continue in STEM fields throughout college and in the workforce. This will be accomplished by reviewing five themes in the literature that are both internal and external to makerspaces. These include: effective program leaders, changing the narrative of the space, building safe spaces, creating support systems, and creating opportunities for wellness and mental health monitoring. This chapter provides an overview of these themes, emphasizing the critical role of efficacious leadership in school-based makerspaces. This work also explores how a mathematics, making, and social justice graduate course impacts secondary teacher beliefs about incorporating social justice into the mathematics classroom.

INTRODUCTION

Although technical advances are ever accelerating, persons who are Black, Indigenous, and People of Color (BIPOC) continue to be underrepresented in science, technology, engineering, and mathematics (STEM) fields (Funk & Parker, 2018). As a result, artificial intelligences, technological developments, and digital tools espouse racist and white-centric characteristics (Benjamin, 2020; Noble, 2018; Zou & Schiebinger, 2018). To counter these injustices and close the digital gap, makerspaces have been proposed as a tool for increasing participation and training for underrepresented groups in STEM fields (Barton et al., 2017). Unfortunately, makerspaces have not been the easy panacea anticipated and the literature suggests that specific strategies must be employed to increase diverse participation in makerspaces (Barton et al., 2017).

The literature demonstrates that there are several thematic approaches for promoting BIPOC student comfort, engagement, and belonging in makerspaces, including: recruiting successful leadership, creating safety, redefining a space, wellness monitoring, and student support. However, the literature indicates that the five approaches have never been jointly considered in addressing BIPOC student comfort in makerspaces. This chapter argues that these combined approaches are essential to creating comfort and belonging for students and that institutional makerspaces must utilize these approaches to form a comprehensive plan for addressing STEM inclusion. Without institutional change that utilizes these approaches, the STEM gap will continue to grow for BIPOC students, creating an ever more unjust digital society.

HISTORY AND CONTEXT

Makerspaces are defined as modern workspaces that bring together a variety of STEM materials and technologies in schools and communities to tinker, design, learn, and explore. Some technologies are firmly stationed in the space, while others may be modular and can be mobilized and brought to different learning venues. These spaces and technologies are designed for both communal collaboration and individual projects, exposing communities to contemporary tools, such as: computers, Arduinos, three-dimensional (3D) printers, robotics equipment, laser and vinyl cutters, movie-making technologies, and soldering irons (Barton et al., 2017; Blackley et al., 2017; Sheridan et al., 2014). These resources allow members of the space, informally known as "makers," to engage in design thinking, prototyping, and modifying projects with support from local equipment and technological experts, who lead and coordinate the makerspaces. The spaces provide access to technologies, equipment, and expertise that are otherwise costly and inaccessible to local communities, K-12 schools, and global networks of creators and problem-solvers (Blackley et al., 2017; Peppler & Bender, 2013; Sheridan et al., 2014).

The zeitgeist for makerspaces began in the 2000's in conjunction with FabLabs, the rise of Make Magazine, and MakerFaires (Davis, 2015; Dougherty, 2013; Taheri et al., 2019). In 2003, the Massachusetts Institute of Technology bolstered the *How to Make (Almost) Anything* course by including digital fabrication tools such as 3D printers and electronics prototyping equipment in their first Fabrication Lab (FabLab) (Gershenfeld, 2019; Taheri et al., 2019). Although that space had an emphasis on design technologies, it can still be considered one of the first modern makerspaces. Shortly thereafter, Make Community, LLC started a Do-It-Yourself (DIY) community movement through the creation of Make Magazine in 2005 and the first MakerFaire in 2006 (Burke & Kroski, 2018; *Make*, n.d.). These institutions created accessible instructions for a variety of DIY projects that encompassed the wide scope of engineering, fashion, technological, and scientific endeavors.

Since that time, the enthusiasm for hands-on, programmable, and constructible educational manipulatives has driven STEM education in an effort to prepare students for twenty-first-century learning (Sanders, 2009). STEM activities in makerspaces provide a variety of academic benefits, including: practice with design, prototyping, communication, collaboration, teamwork, critical thinking, and academic enrichment (Kiley-Rendon, 2019; Taheri et al., 2019). Although Gilbert (2017) suggests makerspaces may become redundant as robotic manufacturing replaces manual labor, she also acknowledges the enormous potential for makerspaces to increase equal access to technologies and STEM knowledge if levied appropriately. As such, makerspaces should be considered a powerful tool for preparing students for 21st-century STEM careers.

In spite of the prolific nature of makerspaces in schools, BIPOC students are not regular participants in these labs (Kafai & Burke, 2014; Sang & Simpson, 2019) and given the well-documented relationship between behavior and identity (Simons, 2021), makerspaces must be carefully established to foster the belonging that traditionally underrepresented students do not regularly feel in engineering programs (Cirell et al., 2020). This trend continues into the STEM workforce (Jones et al., 2018; Kwasa, 2021; Pourret et al., 2021) and as a consequence, white-centric, racially-insensitive, and inequitable standards become deeply embedded in new technologies, including discriminatory search engine algorithms and white supremacist artificial intelligences (Noble, 2018; Zou & Schiebinger, 2018). An abhorrent example is the 2015 Google facial recognition "glitch" that automatically tagged Black, African American, and People of Color as "gorillas," "apes," and "animals" (Noble, 2018). Furthermore, facial recognition software incorrectly identifies Black, African American, and People of Color ten to a hundred times more often than white faces and has greater difficulty recognizing women in comparison to men (Singer & Metz, 2019). A greater representation of BIPOC computer programmers changes this narrative.

Systems of institutional, medical, societal, and environmental racial oppression have been well-documented in the United States (L. T. Brown, 2021; Kendi & Reynolds, 2020; Washington, 2006, 2020; Wilkerson, 2020), the effects of which have extended into computer science, contemporary technology, and digital invention (Benjamin, 2020). The shortage of Black computer programmers, engineers, mathematicians, and scientists echoes these divisions (Benjamin, 2020; Jones et al., 2018; Pourret et al., 2021). In 2014, just 6% of the STEM workforce was comprised of Black, Latine, and Indigenous employees (Barton et al., 2017), reiterating that makerspaces and STEM fields are dominated by the white, male narrative (Kafai & Burke, 2014; Vossoughi et al., 2016). The majority of women in STEM fields report discrimination and in computer science, female representation has decreased to just 25%

since 1900 (Funk & Parker, 2018). Without measures to address this representation gap by creating a sense of belonging and STEM comfort during formative years, we will continue to see an ever-widening digital divide that leaves our BIPOC populations out of critical STEM industries.

THEMES

The literature suggests a multitude of strategies for combatting a lack of diverse participants in school makerspaces, and these can be grouped into five main categories: effective program leaders, building safe spaces, changing the narrative of makerspaces, monitoring student health, and implementing support systems. These thematic approaches to increasing BIPOC student comfort in makerspaces can be grouped into two camps: internal and external makerspaces factors. Internal factors are outlined as characteristics that impact student participation within the makerspace, specifically the conditions of the space and relationships between students and faculty. External factors exist outside of the makerspace and include obstacles to participation, such as stress and mental health, and academic barriers, including GPA and advanced or honors classes. The interplay of internal and external factors provides a comprehensive approach to boosting student comfort and belonging in makerspaces.

INTERNAL FACTOR: EFFECTIVE PROGRAM LEADERS

Program leaders with high levels of efficacy play a crucial role in successful STEM institutions that have high levels of diverse student representation and low levels of BIPOC student attrition rates. Makerspaces are frequented and frequently managed by white, adult men, whose interests may not align with BIPOC student interests and community needs (Barton et al., 2017). This can cause disinterest and disassociation on the part of BIPOC students (Douglas et al., 2008). It is crucial to employ effective leaders to ensure that makerspaces focus on community priorities, promote strong relationships between mentors and students, and offer expertise in the use of technical tools. These leaders drive the direction of the space, norms, and culture.

Leaders who commit to long-term mentoring relationships and bring technical expertise to the community are a critical component of makerspace success (Barton et al., 2017; Jett & Davis, 2020; Kwasa, 2021; Masters et al., 2018). In the "Making 4 Change" program, Barton, Tan, and Greenberg (2017) explored the characteristics of successful mentors. Program participants included thirty-six Black and Latine K-12 students and the founders sought leadership from the community who were knowledgeable about STEM, tools, and procedures. Mentors were invited to participate in a long-term experience mentoring students in the program. Although it was challenging to find leaders who reflected the diverse participants (women, people of color, etc.), the relationships of the mentors with the students played a key role in student retention and success (Barton et al., 2017). Furthermore, Jett and Davis (2020) demonstrated the need for effective leadership in their literature review on factors that most significantly contribute to the success of Black male students in STEM coursework. The results implied that long-term and dependable relationships with faculty and mentors played a major role in engagement with STEM and continued success. These relationships extended outside of the classroom to extra-curricular STEM experiences. The most significant relationships were those between BIPOC students and Black, male STEM faculty (Jett & Davis, 2020).

Moreover, makerspace leaders must represent and center on the needs of the community (Barton et al., 2017; Masters et al., 2018; Riley et al., 2017). Masters et al. (2018) determined that makerspaces should be driven by leaders who prioritize the needs of local participants as opposed to institutional priorities. Masters et al. (2018) reviewed six makerspaces that were managed by mentors who aimed to recruit specific demographic groups, such as women, People of Color, and other marginalized STEM populations, and acquired technologies or crafted activities that supported their members. For instance, the group Liberating Ourselves Locally in Oakland, California, was initiated and run by people of color and focuses on endeavors for community sustenance, such as cuisine, textiles, accommodations, art, technologies, and culture (Masters et al., 2018). These programs engaged community members, especially people of color, and supported participants in their endeavors within the makerspace.

INTERNAL FACTOR: SAFE SPACES

A second factor impacting BIPOC student participation is perceived physical and emotional safety in a makerspace. Students who do not feel physical and emotional safety in a makerspace are less likely to engage in STEM activities in the space (Bradshaw et al., 2014). Feelings of safety are evidenced by student comfort in testing, tinkering, and sharing their beliefs, questions, and personality with the making community, and engaging in collaborative projects while having the confidence to make creative mistakes. The cultivated culture of the space is non-punitive, and students feel comfortable and capable of working within the constraints of the space (Bradshaw et al., 2014).

One method for promoting feelings of emotional and physical safety in a makerspace is to create spaces that are representative of the shared identities of participants (Barton et al., 2017; Holbert, 2016; Masters et al., 2018; Vossoughi et al., 2016; Young et al., 2013). Makerspaces and STEM activities that target marginalized populations have been shown to yield positive results for recruitment and retention of BIPOC participants. Masters et al. (2018) reviewed several makerspaces that focused on a particular demographic group and their goals; thus creating a feeling of belonging for participants. Two examples that focus on demographic groups include the Liberating Ourselves Locally Makerspace in Oakland and the MergeSort Makerspace in New York. The Liberating Ourselves Locally makerspace is managed and run by people of color and focuses on community enhancement, whereas the MergeSort Makerspace promotes women and non-binary makers in the STEM arena (Masters et al., 2018). Both spaces created an environment where participants feel as though they are part of a community that prioritizes their identity.

An additional approach for making a space safe is to increase comfort with technologies through tinkering, play, and normalized risk taking (Holbert, 2016; Papert, 1991; Vossoughi et al., 2016). Seymour Papert's constructionist approach (1980, 1991) suggested that students are best able to construct knowledge, especially in computer science, by playing with technologies. Holbert's (2016) case study utilizes Papert's approach by creating playful learning opportunities for BIPOC fourth grade children. This "Bots for Tots" program allows students to use makerspace technologies and build toys for younger students. A secondary goal of the program was to bring students' own culture and traditions into the toys. Holbert analyzed data collected in pre- and post-interviews of the nine participants and identified that students enjoyed exploring 3D printing, laser cutting, and fabrication technologies to meet the toy requests of the Pre-K students. The ability to tinker and play with the technologies helped students feel confident in their ability to use the equipment to manufacture the toys (Holbert, 2016). Vossoughi (2016) also recommended that makerspaces normalize risk taking in their practices in order to promote BIPOC student comfort in the space.

INTERNAL FACTOR: CHANGING THE NARRATIVE OF THE MAKERSPACE

Another internal factor in increasing BIPOC student comfort in makerspaces is changing the student perception of a space (Barton et al., 2017; Kafai & Burke, 2014; Sang & Simpson, 2019; Vossoughi et al., 2016). This can be accomplished by adjusting the perception of projects, incorporating civic engagement, and normalizing regular participation in the space. These approaches reframe the space in a way that entices students to learn and tinker without fear of social ostracization.

It is critical for students to perceive that spaces are trendy, cutting edge, and intended for their interests (Sang & Simpson, 2019; Vossoughi et al., 2016). In addition to combatting student perceptions about the developmental appropriateness of a space, makerspaces must also have some degree of a "cool" factor that entices participation (Sang & Simpson, 2019). Youth identities must be reflected in the space in order to allow BIPOC students to feel ownership of the technologies and the room (Barton et al., 2017). Furthermore, BIPOC students may feel alienated by projects that reflect white privilege (Vossoughi et al., 2016) or are inaccessible due to academic requirements (Sang & Simpson, 2019). To maintain interest, comfort, and belonging, special care must be taken when curating technologies and activities for makerspaces.

Moreover, makerspaces should make use of tools for civic engagement, social justice, and sharing cultural identities whenever possible. Culturally responsive design tools, such as ethnocomputing, combine student culture and computer programming (Kafai & Burke, 2014). This kind of project-based learning activity promotes civic engagement and awareness of culture and cultural heritage. Kafai and Burke (2014) used ethnocomputing with Arduino Lilypads to teach indigenous secondary students about programming through e-textile manufacturing of quilts that reflected student interests and cultures. Similarly, Barton, Tan, and Greenberg (2017) reviewed the "Making 4 Change" program where students created jackets that were both warm and well-lit to help prevent violence against women at night. Both projects allowed students to share their identity and provide responses for certain issues. In each case, prolonged participation in the space was a key component of the project's success (Barton et al., 2017; Kafai & Burke, 2014). Each case required a mindset shift, both on the part of the students and from faculty who need to reframe what can be done to engage students.

EXTERNAL FACTOR: WELLNESS AND MENTAL HEALTH MONITORING

Outside of the makerspace, BIPOC students who are involved in STEM programs should have additional wellness and mental health monitoring to assess student stress of STEM participation. These wellness checks should be a regular component of participation and provide students with opportunities to articulate their experiences and reflect on the balance of their stress with STEM activities. Furthermore, therapists who are Black, African American, and People of Color and are paired with clients who identify similarly reported increased feelings of connection and concern for patient well-being (Goode-Cross & Grim, 2016), suggesting the benefits of comparable partnerships for BIPOC STEM students. The impact of race-related trauma and anxiety on Black students' health and wellness has been documented in multiple studies (Adam et al., 2015; Washington, 2020; Wilkerson, 2020). Cortisol levels have been used to measure stress and results indicate that racially charged incidents spike cortisol levels. While cortisol itself does not negatively influence health, it is associated with higher levels of stress, which has been associated with numerous negative implications for wellbeing (Adam et al., 2015; Richman & Jonassaint, 2008; Washington, 2006).

Richman and Jonassaint (2008) found that Black college women who watched clips of civil-rights speeches had cortisol levels that were significantly higher compared with the control group who watched clips of school athletic events. A longitudinal study by Adam, Heissel, and Zeiders (2015) also demonstrated hypercortisolism level changes for Black children in comparison with their white peers that had resulted from long-term, traumatic childhood experiences or stress relating to race (Adam et al., 2015). Both studies demonstrate that while cortisol may not cause health complications, lower levels of cortisol are associated with better health outcomes surrounding stress, cardiac health, and sleep hygiene. The results of both studies are concerning, because prolonged adrenocortical activation is correlated with negative health outcomes, such as increased risk of infectious disease, depression, fibromyalgia, fatigue, and post-traumatic stress disorder (PSTD) (Adam et al., 2015; Richman & Jonassaint, 2008). Although less racially charged than the previous studies, as previously mentioned, STEM experiences can heighten BIPOC student awareness of underrepresentation in a space, which can cause stress (Hall & Newman, 2020).

EXTERNAL FACTOR: SUPPORT SYSTEMS FOR PARTICIPANTS

Another external factor for increasing comfort for BIPOC students in makerspaces is creating support systems for students. These supports are defined as systems that eliminate barriers to makerspace activities, form connections between participants, and supplement student knowledge. Without support systems, students can feel less engaged in programs and leave the program when they are faced with obstacles (Kwasa, 2021).

The most immediate obstacle to student participation in STEM activities is financial barriers limiting materials, program enrollment, and transportation (Kwasa, 2021; Masters et al., 2018). Riley, McNair, and Masters (2017) identified that a resolution to this barrier is the removal of financial hardship via sliding scale memberships, free experiences, or shared memberships. Some of the organizations leased spaces that were easily accessible via public transit or offered exhibits for makers to share and sell their works. Kwasa (2021) found that providing academic assistance for first-generation college scholars and removing financial burdens associated with academia were important supports for BIPOC students in STEM majors. Black families are more likely to have additional financial burdens associated with medical bills (Washington, 2006, 2020), so allowing students an additional opportunity to earn money via campus jobs that pay more than minimum wage helped alleviate student financial stress (Kwasa, 2021).

Additionally, peer and adult support groups yielded positive results in increasing student comfort (Barton et al., 2017; J. Brown et al., 2018; Jett & Davis, 2020). Jett and Davis (2020) found that peer support groups create learning teams and study groups help Black male students achieve and excel in STEM coursework. Extra-curricular or specialized experiences that reinforce STEM content have also been shown to support student success, especially if started at an early age. Smith (2020) also supports the formation of affinity groups to allow individuals to feel a sense of belonging. Hurley Boykin, and Allen (2005) also support the idea of communal learning to connect knowledge in and out of school and removing scheduling conflicts is also important (Sang & Simpson, 2019).

IMPLEMENTING TRAINING TO FOSTER INCLUSIVE MAKING LEADERSHIP

Although all facets of inclusive makerspaces are significant, the most influential catalyst of change is arguably an effective makerspace leader who can formulate a combined approach to create inclusive makerspaces. While institutional pressures and standards provide some scaffolding for these factors, an effective leader ultimately ensures the establishment and consistent implementation of all justice-centered support systems, safety, and engaging spaces. This individual solidifies the standards, expectations, and agenda for the makerspace and subsequent projects and programming. However, honing these leadership skills requires specific training in makerspace technologies, pedagogy, content knowledge, and justice-centered, inclusive learning. Although many teacher preparation programs also focus on building justice-centered learning into STEM classes (Cochran-Smith, 2010), there is limited research on the role of race, ethnicity, culture, and social justice training in mentoring relationships between white mentors and youths of color (Anderson & Sánchez, 2022), especially in STEM settings. However, the research that exists suggests the importance of adult training centering on these issues (Anderson & Sánchez, 2022). Given the volatile and sensitive nature of discussing these topics in the classroom, it is more important than ever that teachers and mentors for STEM learning receive adequate training in order to foster inclusive makerspace leaders.

Studies show that mentoring relationships with adequate foundational training on justice-centered and cultural issues yield greater success between white mentors and students of color (Anderson & Sánchez, 2022; Henneberger et al., 2013; McGill, 2012). In comparison with a control group, Anderson and Sánchez (2022) demonstrated that training mentors in racial, ethnic, cultural, and social justice topics impacts their relationship satisfaction and quality of interventions by improving mentor self-efficacy in providing culturally-sensitive support for their mentees. The mentors who received training were better able to support and validate youths after becoming more confident in their own ability to identify systemic privilege and oppression. Similarly, the Young Women Leaders Program (YWLP) at the University of Virginia also provides training for their predominantly white mentors in supporting mentees of color to grow their skills in connection and autonomy (Henneberger et al., 2013). Students participating in YWLP were shown to maintain global self-esteem over time in the program in comparison with the declines of self-esteem for the control group. These studies demonstrate the importance of providing adequate training in inclusion-and-justice-centered topics to promote sustained and effective leadership in inclusive makerspaces.

OUR RESEARCH STUDY

Although the difficulties of changing teacher beliefs surrounding social justice topics are well-documented, teacher preparation programs that implement specific justice-centered learning training can encourage teachers to examine their beliefs, discuss and challenge inequity, and incorporate teaching practices that celebrate diversity (Enterline et al., 2008). However, the literature suggests that additional research demonstrating the efficacy of justice-centered training in STEM teacher training programs is needed (Anderson & Sánchez, 2022). Our work focuses on the students in a mathematics, making, and social justice graduate course who experienced a change in beliefs about the role of social justice in mathematics classes following targeted social justice and mathematical-making training. These students are secondary teachers who are enrolled in a Mathematics Education M.S. degree program. Seven of the enrolled teachers elected to participate in a post-course survey reflecting on how their beliefs changed throughout the course. For more information about the course, participants, and survey, please see the chapter "Preparing STEM Teachers to be Change Makers" in this book. Teachers who participated in the post-class survey self-reported little previous experience with integrating social justice into their STEM lessons but shared enthusiasm for future incorporation of social justice via STEM activities after the interventions.

Each social justice training intervention followed the following formula:

- Demonstrate the need for social justice topics including accessibility, water loss, lack of access to nutritious food, the need for a celebration of diverse heroes, and climate change.

- Model the implementation of the social justice content complementing the technology, makerspace pedagogy, and STEM content.

- Reflect on the experience and challenges of implementing social justice and making.

- Engage in opportunities to practice creating social justice STEM lessons and projects for their students.

The framework for the activity design followed an expansion of Mishra and Koehler's (2008) Technology, Pedagogical, and Content Knowledge (TPACK) framework that also incorporated making and inclusion as proposed in the Inclusive MakerPACK framework (Corum et al., 2020; Nichols & Corum, 2023).

RESULTS AND DISCUSSION

Although not all teachers participated in the survey, several included enthusiastic comments that indicated their change in beliefs about the importance of social justice. The survey collected short responses to questions about changing social justice beliefs and results were analyzed using an inductive and deductive coding approach (Saldaña, 2021).

One teacher walked away from the training experiences feeling an increased sense of obligation to take civic action or implement social justice in their STEM classes as indicated in the response below.

> "...It made me realize that we need to take our earth seriously. The second lesson made me realize how fortunate and blessed I am with having many options for shopping. It also made me curious about how many communities do not have [nutritious grocery store food] and how to help change that."

The training and activities brought awareness to the social justice issues embedded in climate change and the lingering detrimental impact of historic redlining in urban environments. Anderson and Sánchez (2022) argue that this awareness and empathy have positive impacts on mentor-mentee relationships who have different racial and cultural backgrounds. Another teacher acknowledged the synergy between STEM, maker technologies, and social justice and enthusiasm for solving equity-based problems following the training:

> "I found the social justice-centered lessons inspire to help bring awareness to problems in our world. We need solutions. We need engaged problem solvers using all they can. Using tech help brings

a different perspective and offers an additional layer of seeing and solving problems."

Other teachers reiterated their excitement for bringing social justice into a mathematics classroom. One teacher noted the importance of being aware of social justice considerations when presenting a mathematics lesson:

"Bringing social issues to light in a teaching environment sparked new ideas for presenting math lessons in an engaging and socially aware manner."

This demonstrated increased awareness of social and cultural considerations. Anderson and Sánchez (2022) reiterate the importance of this cognizance, as well-meaning mentors who lack cultural awareness may offend mentees and lead to the early termination of relationships. Similarly, another teacher emphasized the importance of including social justice activities in mathematics class on a regular basis:

"I believe that social justice can and should be brought into math lessons daily/weekly because students already have such a difficult time realizing the importance of math in real life."

This indicates the realization that social justice is an important component of learning in STEM environments and that STEM teachers have ownership over this implementation. As a result of the training that combines STEM, makerspace technologies, and social justice, these teachers are empowered to incorporate justice-centered making in their mathematics classrooms.

CONCLUSION

This work demonstrates the impact of social justice STEM training on creating confidence and self-efficacy for effective leaders in creating sustainable relationships with BIPOC youths in makerspaces. While much literature exists on the importance of social justice in STEM curricula, this research suggests that specific justice-centered maker training opportunities yield leaders who have the awareness and skills to create makerspace programming that is effective and inclusive. Although the sample size is small, the data yields positive initial results and a variety of opportunities for future research into STEM, social justice, and makerspace leadership. This work has future implications for subsequent study of the teachers' involvement in technology leadership, use of makerspace technologies in the classroom, and experiences engaging BIPOC students in STEM work. An additional area of study can explore whether recruiting successful leadership, creating safety, redefining a space, wellness monitoring, and student support impact BIPOC student participation in Makerspaces. In an increasingly complex world, it is crucial that teachers act as leaders and change agents in engaging BIPOC students in STEM.

REFERENCES

Adam, E. K., Heissel, J. A., Zeiders, K. H., Richeson, J. A., Ross, E. C., Ehrlich, K. B., Levy, D. J., Kemeny, M., Brodish, A. B., Malanchuk, O., Peck, S. C., Fuller-Rowell, T. E., & Eccles, J. S. (2015). Developmental histories of perceived racial discrimination and diurnal cortisol profiles in adulthood: A 20-year prospective study. *Psychoneuroendocrinology*, *62*, 279–291. https://doi.org/10.1016/j.psyneuen.2015.08.018

Anderson, A. J., & Sánchez, B. (2022). A pilot evaluation of a social justice and race equity training for volunteer mentors. *American Journal of Community Psychology*, *69*(1-2), 3–17. https://doi.org/10.1002/ajcp.12541

Barton, A. C., Tan, E., & Greenberg, D. (2017). The makerspace movement: Sites of possibilities for equitable opportunities to engage underrepresented youth in STEM. *Teachers College Record: The Voice of Scholarship in Education*, *119*(6), 1–44. https://doi.org/10.1177/016146811711900608

Benjamin, R. (2020). Race after technology: Abolitionist tools for the new Jim Code. *Social Forces, 98*(4), 1–3. https://doi.org/10.1093/sf/soz162

Blackley, S., Sheffield, R., Maynard, N., Koul, R., Walker, R. (2017). Makerspace and reflective practice: Advancing pre-service teachers in STEM education. *Australian Journal of Teacher Education, 42*(3), 22–37. https://doi.org/10.14221/ajte.2017v42n3.2

Bradshaw, C. P., Waasdorp, T. E., Debnam, K. J., & Johnson, S. L. (2014). Measuring school climate in high schools: A focus on safety, engagement, and the environment. *Journal of School Health, 84*(9), 593–604. https://doi.org/10.1111/josh.12186

Brown, J., Schreiber, C., & Barbarin, O. (2018). Culturally competent mathematics instruction for African American children. In M. Caspe, T. A. Woods, & J. L. Kennedy (Eds.), *Promising practices for engaging families in STEM learning: A volume in family school community partnership issues* (pp. 49–61). Information Age Publishing, Inc.

Brown, L. T. (2021). *The black butterfly: The harmful politics of race and space in America*. Johns Hopkins University Press.

Burke, J., & Kroski, E. (2018). *Makerspaces: A practical guide for librarians* (2nd ed.). Rowman & Littlefield.

Cirell, A. M., Kellam, N., Boklage, A., & Coley, B. (2020). Reimagining the thirdspace through makerspace. In M. Melo & J. T. Nichols (Eds.), *Re-making the library makerspace: Critical theories, reflections, and practices* (pp. 47–82). Library Juice Press.

Cochran-Smith, M. (2010). Toward a theory of teacher education for social justice. In A. Hargreaves, A. Lieberman, M. Fullan, & D. Hopkins (Eds.), *Second International Handbook of Educational Change* (pp. 445–467). Springer Netherlands. https://doi.org/10.1007/978-90-481-2660-6_27

Corum, K., Nichols, L., Spitzer, S., & Begen, K. (2020). Supporting teachers' ability to leverage makerspaces in the teaching and learning of mathematics. In D. Schmidt-Crawford (Ed.), *Proceedings of Society for Information Technology & Teacher Education International*

Conference (pp. 1334–1339). Online: Association for the Advancement of Computing in Education (AACE). https://www.learntechlib.org/primary/p/216043/

Davis, M. (2015, December 10). A brief history of makerspaces. *Curiosity Commons.* https://curiositycommons.wordpress.com/a-brief-history-of-makerspaces/

Dougherty, D. (2013). The maker movement. In M. Honey & D. Kanter (Eds.), *Design, make, play: Growing the next generation of STEM innovators* (pp. 7–11). Routledge.

Douglas, B., Lewis, C. W., Douglas, A., Scott, M. E., & Garrison-Wade, D. (2008). The impact of White teachers on the academic achievement of Black students: An exploratory qualitative analysis. *Educational Foundations, 22,* 47–62.

Enterline, S., Cochran-Smith, M., Ludlow, L. H., & Mitescu, E. (2008). Learning to teach for social justice: Measuring change in the beliefs of teacher candidates. *The New Educator, 4*(4), 267–290. https://doi.org/10.1080/15476880802430361

Funk, C., & Parker, K. (2018). *Women and men in STEM often at odds over workplace equity.* Pew Research Center. https://www.pewresearch.org/social-trends/2018/01/09/diversity-in-the-stem-workforce-varies-widely-across-jobs/

Gershenfeld, N. (2019). *Making (almost) anything: Neil Gershenfeld outlines the impact of fab labs.* MIT Spectrum. https://spectrum.mit.edu/fall-2019/making-almost-anything/

Gilbert, J. (2017). Educational makerspaces: Disruptive, educative or neither? *New Zealand Journal of Teachers' Work, 14*(2), 80–98.

Goode-Cross, D. T., & Grim, K. A. (2016). "An unspoken level of comfort": Black therapists' experiences working with Black clients. *Journal of Black Psychology, 42*(1), 29–53. https://doi.org/10.1177/0095798414552103

Hall, K., & Newman, C. B. (2020). Ease on down the road: Navigating the Yellow Brick Road to graduation at a Primarily White Institution (PWI). In C. S. Platt, A. A. Hilton, C. B. Newman, & B. Hinnant-Crawford (Eds.), *Multiculturalism in higher education: Increasing access and improving equity in the 21st century* (pp. 109–128). Information Age Publishing, Inc.

Henneberger, A. K., Deutsch, N. L., Lawrence, E. C., & Sovik-Johnston, A. (2013). The Young Women Leaders Program: A mentoring program targeted toward adolescent girls. *School Mental Health, 5*(3), 132–143. https://doi.org/10.1007/s12310-012-9093-x

Holbert, N. (2016). Leveraging cultural values and "ways of knowing" to increase diversity in maker activities. *International Journal of Child-Computer Interaction, 9–10*, 33–39. https://doi.org/10.1016/j.ijcci.2016.10.002

Jett, C. C., & Davis, J. (2020). Black males' STEM experiences: Factors that contribute to their success. In E. O. McGee & W. H. Robinson (Eds.), *Diversifying STEM: Multidisciplinary perspectives on race and gender* (pp. 192–208). Rutgers University Press.

Jones, J., Williams, A., Whitaker, S., Yingling, S., Inkelas, K., & Gates, J. (2018). Call to action: Data, diversity, and STEM education. *Change, 50*(2), 40–47. https://doi.org/10.1080/00091383.2018.1483176

Kafai, Y. B., & Burke, Q. (2014). Connected gaming: Towards integrating instructionist and constructionist approaches in K-12 gaming. In J. L. Polman, E. A. Kyza, K. O'Neill, I. Tabak, W. R. Penuel, A. Susan. Jurow, K. O'Connor, & L. D'Amico (Eds.), *Learning and Becoming in Practice: The International Conference of the Learning Sciences (ICLS)* (Vol. 1, pp. 86–93). International Society of the Learning Sciences.

Kendi, I. X., & Reynolds, J. (2020). *Stamped: Racism, antiracism, and you* (1st ed.). Little, Brown and Company.

Kiley-Rendon, P. A. (2019). *The academic benefits of makerspaces for middle-childhood students*. ProQuest LLC. https://www.proquest.com/docview/2268372929

Kwasa, J. A. C. (2021). Paying attention to what matters: Individual differences in neural correlates of spatial selective attention in heterogeneous populations and increasing diversity, inclusion, equity, and justice in STEM [Ph.D., Carnegie Mellon University]. In *ProQuest Dissertations and Theses*. http://www.proquest.com/docview/2543455076/abstract/5D2287B96FED4D39PQ/1

Make: Community. (n.d.). Make: Community. Retrieved October 9, 2021, from https://make.co

Masters, A. S., McNair, L. D., & Riley, D. M. (2018). MAKER: Identifying practices of inclusion in maker and hacker spaces with diverse participation. *Proceedings of the ASEE Annual Conference & Exposition*, 1–8.

McGill, J. M. (2012). *The Young Women Leaders' Program: Exploring factors and outcomes associated with emerging adult female mentors' experience* [Auburn University]. https://etd.auburn.edu/xmlui/bitstream/handle/10415/3398/McGill%20Thesis%20Final.pdf?sequence=2&isAllowed=y

Mishra, P., & Koehler, M. J. (2008). *Introducing technological pedagogical content knowledge*. 1–16.

Nichols, L., & Corum, K. C. (2023). Increasing teacher commitment to justice-centered mathematics through maker-enhanced social justice activities. In E. Langran, P. Christensen, & J. Sanson (Eds.), *Society for Information Technology & Teacher Education International Conference* (pp. 411–416). Association for the Advancement of Computing in Education (AACE).

Noble, S. U. (2018). *Algorithms of oppression: How search engines reinforce racism*. NYU Press. https://doi.org/10.2307/j.ctt1pwt9w5

Papert, S. (1980). *Mindstorms: Children, computers, and powerful ideas*. Basic Books.

Papert, S. (1991). Situating constructionism. In S. Papert & I. Harel (Eds.), *Constructionism* (pp. 1–11). Ablex.

Peppler, K., & Bender, S. (2013). Maker movement spreads innovation one project at a time. *Phi Delta Kappan*, *95*(3), 22–27. https://doi.org/10.1177/003172171309500306

Pourret, O., Anand, P., Arndt, S., Bots, P., Dosseto, A., Li, Z., Marin Carbonne, J., Middleton, J., Ngwenya, B., & Riches, A. J. V. (2021). Diversity, equity, and inclusion: Tackling under-representation and recognition of talents in geochemistry and cosmochemistry. *Geochimica et Cosmochimica Acta*, *310*, 363–371. https://doi.org/10.1016/j.gca.2021.05.054

Richman, L. S., & Jonassaint, C. (2008). The effects of face-related stress on cortisol reactivity in the laboratory: Implications of the Duke lacrosse scandal. *Annals of Behavioral Medicine*, *35*(1), 105–110. https://doi.org/10.1007/s12160-007-9013-8

Riley, D. M., McNair, L. D., & Masters, S. (2017). An ethnography of maker and hacker spaces achieving diverse participation. *Proceedings of the ASEE Annual Conference & Exposition*, 5952–5956.

Saldaña, J. (2021). *The coding manual for qualitative researchers* (4th ed). SAGE Publishing.

Sanders, M. (2009). STEM, STEM education, STEMmania. *The Technology Teacher*, *68*(4), 20–26.

Sang, W., & Simpson, A. (2019). The maker movement: A global movement for educational change. *International Journal of Science and Mathematics Education*, *17*(S1), 65–83. https://doi.org/10.1007/s10763-019-09960-9

Sheridan, K., Halverson, E. R., Litts, B., Brahms, L., Jacobs-Priebe, L., & Owens, T. (2014). Learning in the making: A comparative case study of three makerspaces. *Harvard Educational Review*, *84*(4), 505–531. https://doi.org/10.17763/haer.84.4.brr34733723j648u

Simons, J. D. (2021). From identity to enaction: Identity behavior theory. *Frontiers in Psychology*, *12*, 679490. https://doi.org/10.3389/fpsyg.2021.679490

Singer, N., & Metz, C. (2019, December 19). Many facial-recognition systems are biased, says U.S. study. *The New York Times.* https://www.nytimes.com/2019/12/19/technology/facial-recognition-bias.html

Smith, D. A. J. (2020). *Theoretical framework of inclusiveness at workplace. 7*(1), 11.

Taheri, P., Robbins, P., & Maalej, S. (2019). Makerspaces in first-year engineering education. *Education Sciences, 10*(1), 8. https://doi.org/10.3390/educsci10010008

Vossoughi, S., Hooper, P., & Escudé, M. (2016). Making through the lens of culture and power: Toward transformative visions for educational equity. *Harvard Educational Review, 86,* 206–232. https://doi.org/10.17763/0017-8055.86.2.206

Washington, H. A. (2006). *Medical apartheid: The dark history of medical experimentation on Black Americans from colonial times to the present.* Harlem Moon.

Washington, H. A. (2020). *A terrible thing to waste: Environmental racism and its assault on the American mind.*

Wilkerson, I. (2020). *Caste: The origins of our discontents.* Random House.

Young, J., Young, J., & Hamilton, C. (2013). Culturally relevant project-based learning for STEM education. In M. M. Capraro, R. M. Capraro, & C. W. Lewis (Eds.), *Improving urban schools: Equity and access in K-12 STEM education for all students.* IAP, Information Age Publishing, Inc.

Zou, J., & Schiebinger, L. (2018). AI can be sexist and racist—It's time to make it fair. *Nature, 559*(7714), 324–326. https://doi.org/10.1038/d41586-018-05707-8

This program is supported in part by a grant to Virginia Tech from the Howard Hughes Medical Institute through the Inclusive Excellence Grant.

CHAPTER 11.

CREATING A SPACE IN THE CURRICULUM FOR EFFECTIVE MENTORING TO FOSTER STUDENT CONNECTIONS AND AGENCY

CYNTHIA A. DEBOY; PATRICE E. MOSS; AND KAITLIN R. WELLENS

Vignettes by Mia Ray and Anette Casiano-Negroni

Figure by Adriana Pino-Delgado

ABSTRACT

Mentorship in an undergraduate institution has a significant impact on student success. Inclusion and community are created when students are able to share their cultural, social, and academic experiences with peers and faculty. An initial survey of students in science, technology, engineering, and mathematics (STEM) disciplines at Trinity Washington University demonstrated a need for increased engagement, motivation, and community among our students. To that end, we designed and implemented a sequence of four required one-credit courses embedded within the STEM curriculum. These Mentor Moments support social and emotional wellness, with an emphasis on curriculum and career preparation. Effective embedded mentoring requires a safe space for students to form connections with peers, faculty, and the STEM community members, creating multi-tiered mentoring opportunities. Based on assessments and interactions with our students, we highlight here some essential components of effective mentoring and provide examples of how these elements may be incorporated in dedicated courses and other settings. The implementation of Mentor Moments and companion activities correlate with increased persistence of STEM majors and the desire to pursue STEM careers. More importantly, we experience a spirit within our academic community that lends itself to freedom, belonging, and power in position and voice for women of color in STEM. Through this network, we are removing barriers and increasing opportunities, rooted in authentic relationships.

INTRODUCTION

Excitement and nerves echo through the halls of Trinity Washington University (Trinity) as a new academic year commences. Fresh notebooks opening and soft taps on keyboards signal that our students are ready to take on the workload of a new semester. While sitting quietly awaiting the start of class, thoughts of fear, anxiety, excitement, confusion, anticipation and pride circle in the minds of our students. "What if I fail my classes? What if I am the only person in the class who doesn't get what's going on? What if I don't belong here? How will I make friends? No one in my family has gone to college, how will I know what to do?"

Many students experience these thoughts and emotions as they contemplate the monumental occasion of starting their first day as an undergraduate student. More specifically, these feelings are heightened in students majoring in STEM disciplines, especially marginalized populations, such as women and students of color (Rodriguez et al, 2021; Ong et al., 2018).

> **Box 1: Trinity Statistics**
>
> Trinity's College of Arts & Sciences is the full-time undergraduate historic women's college which is home to science programs including biology, biochemistry, chemistry and forensic science. Greater than 90% of students in Trinity's College of Arts and Sciences are African American and Latina and 80% are first generation college students. Recent funding opportunities from TheDream.US has increased the percentage of students with varying immigration challenges. Trinity is the only institution in D.C., and one of few in the country, to be classified as both a Predominantly Black Institution and a Hispanic Serving Institution by the U.S. Department of Education. Approximately 70% of students receive Pell Grants and have a median family income of $25,000.

Trinity students are 100% female and fabulous, from multiple walks of life. (See Box 1). They are excited, enthusiastic, and eager to morph into the best versions of themselves while in pursuit of their undergraduate STEM degree. One of the key factors impacting student success and belonging in STEM is intentional support through mentorship (Kendricks et al., 2013). Inclusion and community are created when students are able to share their cultural, social, and academic experiences with peers and faculty (Estrada et al., 2018). To build an effective mentorship structure at Trinity, we determined we needed to build this into the curriculum so all our students would benefit from mentorship without sacrificing time for outside responsibilities. To that end, we designed four required one-credit courses embedded within the STEM curriculum to support social and emotional wellness, with an emphasis on curriculum and career preparation. (See Figure 11.1). All four mentoring courses run concurrently with students from all science majors taking the courses together. This was strategically designed for us to meet periodically throughout the semester as one large multi-generational community of women in STEM. We also take this time to break into *Mentor Streams*— one STEM faculty member mentoring a group of students from various stages in the STEM program.

Embedding effective mentoring into the curriculum requires creating a safe space, which is necessary for students to form connections with peers, faculty and members of the STEM community. This creates opportunities for multi-tiered mentoring.

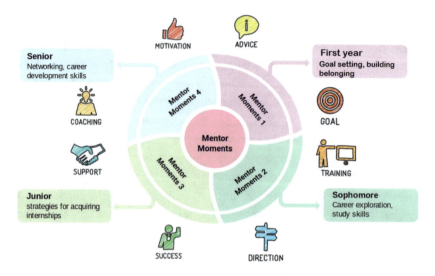

Figure 11.1: Topics and structure for Mentor Moment courses. Figure created by Adriana Pino-Delgado.

MENTORING REQUIRES VALUING STUDENTS' UNIQUENESS AND CELEBRATING DIFFERENCES.

Creating a safe space means developing a community in which all students' uniqueness is valued and differences are celebrated, therefore creating an inclusive environment. Throughout the Mentor Moment courses, students and professors participate in activities in which they learn about and appreciate the uniqueness and value that each brings to our community. In the Mentor Moment 1 course, our first-year science students create vision boards on which they depict their goals and aspirations. They then share their vision boards with students from all mentor classes and STEM faculty during a large group class session. Here, students share and listen to one another's stories and recognize both the unity and diversity of their interests, personalities, and goals. As one student explained, *"I didn't see any pictures that looked like me, so I made one. I added a lab coat with a Trinity ID and a microscope."* As this student so clearly noted, the vision board activity invites students to bring their authentic selves into our community space. As a community, we begin to create a safe space by validating student goals, celebrating who students are as individuals, and fostering a sense of belonging and an awareness that students are not alone.

Students continue to explore career interests and share their goals throughout their second year Mentor Moment course. They develop plans to support their aspirations which they present to peers. Presentations include resources and strategies such as links and deadlines for internship applications. Students develop confidence and leadership as they acquire and share practical information. By inspiring one another and learning together, connections within the community form. As the professor also learns about each student's individual goals, it provides opportunities for future discussions and follow up with specific resources and opportunities. One student noted at the end of the course that she received "mentorship this semester by being reassured" that her chosen major aligns with her career goals.

Mentor Moment courses provide a structure for professors to get to know their students, but more importantly, it provides an opportunity for instructors to communicate to students that they care about their well-being as well as their academic and career goals. In the beginning of their second Mentor Moment course, most Trinity STEM students surveyed (62%) indicated some hesitation about communicating with professors. Student responses to a survey indicated that they hesitate to communicate with professors because they feel intimidated, are afraid of being judged, or worried about bothering their professors. Some students expressed concern that they would not be able to put their question into words clearly. One student explained, *"I hesitate to communicate with my professors most of the time because I get nervous about the reaction they might have to what I'm communicating as well as the kind of response they might have."* The concerns students share help us recognize that, as professors, our reactions to students are critical to encouraging them to open up and seek assistance, but they also have the potential to shut down their efforts. The responsibility therefore falls on us as instructors to create a safe environment for communication, which means that our responses to students need to be affirming and non-judgmental. At the end of the second Mentor Moment course, a student reflected that, *"I think what makes [seeking help] comfortable is how open my professors are to help me. They don't seem bothered by my questions so I feel like I can go to them whenever I have questions".* When our responses to students communicate acceptance of where they are, as well as the high expectations we have for their continued academic growth, we begin to break down barriers perpetuating student hesitation and create an opening for trust and connections that lead to mentorship. (See Student Vignettes 1 & 2.) Through active and responsive listening, we empower students to use their voice for self-efficacy and for learning.

The invitation for students to interact with us as faculty begins with our own understanding of our positionality and recognition of the hesitation students may have communicating with us because of that lens. It has therefore been helpful to provide opportunities for instructors to be more relatable to students by providing examples of overcoming challenges or sharing snippets of our own lives to be more relatable (See Faculty Vignette 1).

Our faculty team participated in a series of training opportunities to learn about topics ranging from implicit bias to culturally affirming assignments to effective mentoring. (For examples of workshops, see Table 11.1.) Each of us incorporated aspects of our approaches to students both formally in our course policies and activities, but also through our awareness and intentionality for every interaction we have with students, recognizing that it all matters. The impact of faculty interactions with students has been both magnified and buffered by the effect of a team approach. This means that students on the whole receive consistent messaging about the value they bring to STEM. Each faculty member communicates this in their own unique interaction with students. These interactions occur not only in the classrooms, but also in the hallways, during office hours (which many of us have renamed student hours), club activities, and science laboratories. Throughout our interactions, and in our own way, we affirm to students their belonging in STEM. We offer our encouragement, motivation, reassurance, and career advice, but this positive interaction only occurs if students feel safe and welcome to open up to our willingness to mentor them. Students have reflected on the value of these interactions. One student indicated, *"I have the support of my professors and the people around me[.] I would say that has a huge impact on how I am performing during this* semester. *They give me advice, help me with any matters and they are also understand[ing].*" Another student added, *"Even if it's just small tips/ advice, I believe it has contributed to my life. There is a professor that has motivated me to keep pushing through in my classes, and I believe that it has aided me a lot mentally and academically.*" The value of mentorship is clear, but the effort and intentionality that faculty must take to welcome students into a space to be vulnerable and trusting enough to be open to mentorship cannot be underestimated.

Table 11.1: Examples of workshops supporting mentorship in the curriculum

Topic	Facilitator	Key takeaways	Outcomes
Cultural competency and universal design	Carl Moore	Inclusive learning environments and practices; value all students	Revised course requirements including mentor classes and interdisciplinary approach
Motivating students to learn and culturally responsive teaching	Christine Harrington	Inclusive syllabi, fostering growth mindset, culturally relevant assignments, meaningful feedback	Addition of program goals including academic and social integration and career exploration
Equity and community building	Bryan Dewsbury	Recognizing privilege	Developing awareness of positionality in mentoring/ teaching
Students become pedagogical partners with faculty	Floyd Cheung	Empowering students in the learning process	Embedded tutor program
Confronting microaggression and building inclusion	Tasha D'Souza	Combatting microaggression and communication skills	Strategies for communication incorporated in Mentor classes
Mentoring students for success	Brett Woods	Increasing student use of support services; supporting students to use active learning strategies to study; peer and faculty mentorship for students; fostering growth mindset	Activities in mentor classes added to foster students' effective use of supports.
Inclusive teaching online: removing barriers in our teaching	Lindsey Masland	Welcoming and engaging students online with check ins and collaborative group activities	Online strategies used to engage students
Optimizing & creating inclusive undergraduate research experiences for students & mentors	Janet Branshaw	Entering Mentoring resource library: http://Cimerproject.org/	Designing mentor courses and developing mentorship skills; preparing students for research experiences

Topic	Facilitator	Key takeaways	Outcomes
Cultural responsive teaching: policy and small group management	Courtney Plotts	Incorporating various voices, historical considerations and perspectives with knowledge of ethnicity and intersectionality. Maximizing ALL students' opportunities to think, socialize, and learn.	Specific activities within courses

NORMALIZING STRUGGLE BUILDS TRUST AND INCREASES ACCESSIBILITY TO CREATE A SAFE SPACE FOR MENTORSHIP

According to Merriam-Webster, The definition of struggle is "to proceed with difficulty or with great effort" however, in our STEM program, we aim to emphasize the process of achieving goals despite difficulty and through struggle. Our students are diverse in many ways and the ability to create an environment where they know they are worth the struggle, for a lack of better terms, is paramount to their mindset and success.

As we meet in our Mentor Streams throughout the semester, we place an intentional emphasis on normalizing struggle. Students and faculty alike share experiences with each other related to issues in STEM and beyond. Students gain the opportunity to learn from each other, citing similarities and differences in their experiences and therefore are able to offer real time and relatable advice, suggestions, and solutions. A student in the sophomore class stated, *"I have had help from other people who also go through the same struggles and can lean on them to know what to do in times of struggle."* These Mentor Streams are also used to humanize faculty through moments of transparency and vulnerability. (See Faculty Vignette 1.) Collectively, this environment removes barriers between students as well as between students and faculty, building trust, increasing accessibility, and creating a safe space.

Our STEM faculty community has most definitely accepted the charge of creating an inclusive environment for the success of our students, but it is important to note that this environment has also been beneficial for the faculty. In the archaic times of our program, before the curriculum redesign, our faculty's interaction with students was minimal. There were some instances where faculty would not meet students until that student's junior or senior year. By this time, the window to form mutual and genuine relationships was significantly smaller. Currently, students have the opportunity to interact with all of our faculty as early as the first semester of their first year. Additionally, the STEM professors that teach our Mentor Moment courses are not necessarily the professors that are instructing the students in their current core STEM courses that semester. Additionally, the Mentor Streams provide exposure to faculty outside of the curriculum year and specific discipline (Biology, Chemistry, Environmental Science, etc.) This structure allows students and faculty to see each other outside of the "science" and get to know each other on a humanistic level, thereby leading to authentic connections and rapport.

The mentoring structure as part of the curriculum has been an effective way to empower and engage students and faculty. However, the efforts to normalize struggle do not end there. It has been shown that introductory STEM courses, including math, can be barriers to student success, especially students from marginalized populations (Chang et al., 2008). Research also indicates that peer to peer tutoring is a great source of support for both the tutor and the tutee (Cutright & Evans, 2016; Rockinson-Szapkiw et al., 2021). Therefore, we decided to create an embedded tutor program for some of the first and second year courses in the curriculum. In short, students who have successfully completed an introductory STEM course serve as tutors, who are incorporated into the classroom space (one tutor per class). These are not tutors who meet after hours or in an isolated space outside of class. They are available for the students enrolled in the course in real time and assist the professor in communicating and demystifying key concepts of the course material. With this program, struggle is being normalized as the advanced students use their voice to empower the learning of their peers; students are leading the charge in learning and mentoring. We have seen some significant gains from the program and have expanded this concept throughout multiple programs at the institution.

FUN ALLOWS STUDENTS TO CONNECT WITH EACH OTHER AND PROVIDE GREATER PEER SUPPORT

Let's get that dopamine flowing! Research shows that dopamine release can lead to memory stimulation (Jay, 2003), meaning that students that have more fun may also be learning more (Bonde et. al., 2014). As part of our mentorship classes, we have specifically allotted time within the curriculum for students to have fun, relax, and get to know each other. We promote conversations that get students to truly know *who* they and their peers are, and what their motivations, interests, and struggles are. These types of conversations and focused time for connection has led to more open peer relationships and true friendships that last through their undergraduate careers. Student reflection data demonstrate how these mentorship opportunities lead to friendships and meaningful relationships: *"I feel I had mentorship this semester because I [got] to bond and get advice from a lot of upper-class students"*. Some examples of fun activities that have helped create a space for student connection and friendships include collaborative games where students work in groups to build the tallest spaghetti tower possible within a given time. Typically, the classroom starts out serious, as students get down to business to determine their best architectural and engineering designs. However, the difficult building materials of spaghetti and tape eventually lead to bursts of laughter and cheering each other on. While the goal of this activity is to realize the power and learning opportunities in one's willingness to try and not succeed, building the spaghetti towers also helps students meet one of the core objectives for the course: to work in teams in a fun and low stakes manner.

As previously discussed, students build vision boards to help them envision their goals and futures in STEM and increase their sense of belonging. During the making of the vision boards, the classroom turns into a lively arts and crafts hub, with music playing and, students running around grabbing magazine clippings, chatting with one another about their designs and their futures. You can hear students have "ah-ha" moments as they recognize their own goals in their peer's dreams and relax into the comfortable space of togetherness.

Building fun and time to play into our mentor courses has set the foundation for peer interactions in their STEM courses. Similarly, the increase in connection and normalizing struggle that occurs between the faculty and students in the mentor moment courses has created a comfortable and relaxed environment that allows for more fun while learning. Furthermore, faculty have incorporated fun into their STEM courses via games or challenges for students. These activities engage students in a hands-on way that also gets them moving, connecting, and having fun while learning. For example, a biochemistry course uses rap battles as a creative way to have students work through concepts. Students work in groups to understand the biochemistry and create the winning rap. The process itself is unique and fun for students to take part in, and the actual battle has students dancing, rapping, cheering, and ultimately connecting through laughter. Overall, by intentionally incorporating fun into the classroom, we are helping our students relax, form a community of support, and be more open to learning.

CELEBRATION FOSTERS TOGETHERNESS AND PRIDE

Celebration is contagious and it is a bug we are happy to spread! We believe that when our students celebrate each other's accomplishments they will use this energy to fuel their own achievements. Furthermore, our STEM students work hard and we want them to feel pride in that work. Therefore, we have built a variety of celebrations, from small, in-class celebrations, to larger university wide celebrations. Growth mindset has been an important core focus of our inclusive pedagogy, so we are mindful to celebrate the road traveled and the growth along the way, not just the end point.

Our Mentor Streams provide the perfect platform for celebrating students. Because we meet in these large groups throughout the semester, we are able to incorporate celebrations of students' achievements and successes with the entire STEM faculty and student body. In the beginning of each Fall semester, we start by celebrating students' achievements from the summer and previous Spring semester by highlighting student scholarship winners and summer fellowships. This not only creates an uplifting energy for the start of the semester and demonstrates to the students that we value their efforts, but it also helps all STEM students become aware of these opportunities.

Toward the end of each semester, we have two main activities in our Mentor Streams centered around celebration. Our first are alumnae panels. Here, we bring alumnae back to campus to speak to students about their experiences while at Trinity and what they are currently doing. We have done similar panels with external scientists, and while they are inspiring, nothing seems to compare to when it is Trinity alumnae. Alumnae are often met with a flurry of student questions, ranging from how they managed their coursework to how they chose and pursued their current careers. Students have mentioned that the alumnae panels resonate with them because they can see themselves in the panel members. By bringing alumnae back to campus, we are also celebrating them and creating a community that extends beyond current Trinity STEM students.

Our second celebratory activity is to throw an end of the year celebration. In recent years, we have tasked the seniors in Mentor Moment 4 with organizing the event for the first through third year students. The energy that the seniors bring to this assignment helps to end the semester on a high note. Students get to hear and learn from their senior peer mentors all while feeling recognized and celebrated by them. It is very clear from student reflections that one of the largest benefits of the Mentor Streams is their interactions with peers in upper-level courses, whether it is helping them build community, normalizing struggle, or celebrating.

Box 2: Measurable outcomes after implementing curricular changes

Since implementing the mentor structure and corresponding curriculum revisions, some measures we have noted include:

- Based on Likert survey responses, students maintain a high sense of belonging throughout their first Mentor Moment course, even as the excitement of beginning college fades and the reality of struggle sets in.

- We found an increase in in average Likert survey responses for how likely students see themselves with a future in science when comparing students in their first to third Mentor Moment classes.

- A student indicates that she has "More confidence in myself, more sure of the career path I want to follow"

- We observed increased pass rates for our introductory biology course.

- We have found increased retention from first semester science courses to second year courses.

- 100% of our science graduates participate in an experiential learning opportunity.

- We have increased the number of science graduates.

We have expanded our celebration of student work to include an annual campus wide event, Spring Research Day. Undergraduate research and experiential learning experiences are high impact practices that increase persistence of women of color in STEM disciplines. However, the process by which to obtain and be successful in these experiences is not always clear and accessible. Through our mentoring structure, students are supported through the process of acquiring internships during Mentor Moment 3. Additionally, students learn from each other and faculty as they share about their undergraduate research and experiential learning opportunities. (See Faculty Vignette 2.) During Spring Research Day, students present their work in an undergraduate research symposium where they are able to see the work of their peers both in STEM and other disciplines. Spring Research Day is highly attended by faculty and administration, creating a sense of pride for the presenters as we celebrate the research collaborators they have become. As part of this event, students receive certificates for their participation. For non-presenting students, Spring Research Day acts as an inspiration for future opportunities and a place to support their peers.

By celebrating our students in a variety of ways, we help to create a community that supports and uplifts one another. The celebrations bring energy to our classrooms and help students feel external pride for their Trinity community as well as an internal pride for their accomplishments.

———————

CONCLUSION

While for some students, the academic undergraduate journey may begin with fear, apprehension, or low self-confidence, we have found that by creating a mentor structure within the curriculum we foster belongingness and support students' perseverance. Tiered mentorship (faculty and peer mentorship) built into the fabric (structure) of our program, supports students through struggles, and fosters supportive relationships that often lead to professional opportunities. We have also embedded fun and celebration throughout the student's academic and mentored experiences. Through our faculty team approach, structured mentorship courses expand into meaningful mentorship relationships and students develop increased confidence and identity within the science field. (See Box 2 for outcomes and Student Vignettes 1 & 2.)

Lessons Learned

- Creating a safe space open for mentorship, necessitates activities, communication and an environment that welcome students to bring themselves to the community space.
- Intentional and consistent affirming and non-judgmental responses to students break down barriers impeding trust and connections necessary for mentorship.
- Normalizing struggle builds trust and increases accessibility to create a safe space for mentorship.
- Celebration and fun help foster togetherness.
- A faculty team approach is an integral component of the effectiveness of the mentor structure embedded within the curriculum.

VIGNETTES

Faculty Vignette 1: Mia Ray

"Mentor Moment I (MM1) is one of my favorite Fall classes to teach. This course gives me an opportunity to drop the façade of a professor and humanize myself to students that I have yet to teach, but hope to encounter during their Junior or Senior year at Trinity. I do this throughout the semester in several different ways. During the first class we do an ice breaker entitled, "What's in a name?". During this activity I introduce myself as Dr. Ray and share a personal story of how I acquired the name. I also have them introduce themselves to the class and a personal aspect of one of their names. This begins to break down the professor/student wall that separates many students from connecting to their instructors and to one another. Another way in which I help to eliminate student/professor barriers is to help students normalize struggle by sharing struggles that I encountered first as a freshman undergraduate student withdrawing from her first Biology Course and later as a first-year medical school student turned Ph.D. Graduate. This level of transparency allows me to connect to students in ways that I wasn't able to prior to MM1. Another way in which I connect with students during MM1 is when I work with them to complete their vision boards. Students begin this section of the course by creating SMART goals (Lawlor and Hornyak, 2012) based on their career aspirations. During this class I share with them a chart of my own SMART goals to show them that goals can continuously be achieved and evolve. As they create their vision boards based on their career goals, I am creating a vision board of my own, which I share with them. These interactions with my students where I allow myself to be vulnerable, allow them to see me as more than just my degree and position, but as someone that they can feel comfortable reaching out to for advice or support."

Mia Ray raym@trinitydc.edu

Faculty Vignette 2: Anette Casiano Negroni

"As a Professor of Chemistry at Trinity Washington University, there is no greater joy than mentoring and seeing our Trinity students succeed in their endeavors. I have learned that my own experiences as an undergraduate student and graduate student in an environment where I had to learn the language simultaneously as I learned the technical skills is one of the reasons why I am even more invested in motivating and encouraging them to push themselves to greater heights. Teaching the Mentor Moment 3 class has given me the opportunity to be part of students' journey as they prepare to apply for experiential learning experiences. I am particularly proud to watch their progress as they participate in internship programs and present their research projects at conferences. As a firm believer in the statement "You can't become what you can't see", I am thrilled when my students get exposed to bigger and challenging STEM environments and come out with a desire to want to learn more and become what they assume at some point in their thought process was impossible. I certainly do not miss a moment to celebrate their accomplishments."

Anette Casiano Negroni casianonegronia@trinitydc.edu

Student Vignette 1: from senior responses to open response questions

"When I first started at Trinity, I did not see myself as a scientist at all. When working with the other [students] in lab groups, I felt insecure that I did not know much as they did … My junior year in the spring semester, a professor pulled me to the side, and she stated she saw potential in me and all i needed to do was be confident in my work and myself [and] I'll go far. … Now that I understand basic concepts, I am willing to discover more in science and pursue other things that relate to science."

> **Student Vignette 2: from senior responses to open response questions**
>
> "I most definitely see myself more as a scientist from when [I] started freshman year. I really struggled with relating to the term when I was in my sophomore year. During that time, My classes were becoming overwhelming and I didn't see myself as a scientist anymore. However, once I started taking more labs and completing more research presentations, I started to enjoy STEM more. The feeling of completing each presentation and showing my classmates our hard work felt great! I also liked a lot of the labs we completed because it peaked my interest in wanting to work in a laboratory."

REFERENCES

Bonde, M. T., Makransky, G., Wandall, J., Larsen, M. V., Morsing, M., Jarmer, H., & Sommer, M. O. (2014). Improving biotech education through gamified laboratory simulations. *Nature Biotechnology, 32*(7), 694-697. https://doi.org/10.1038/nbt.2955

Chang, M. J., Cerna, O., Han, J., & Sáenz, V. (2008). The contradictory roles of institutional status in retaining underrepresented minorities in biomedical and behavioral science majors. *The Review of Higher Education, 31*(4), 433–464.

Cutright, T. J., & Evans, E. (2016). Year-long peer mentoring activity to enhance the retention of freshmen STEM students in a NSF scholarship program. *Mentoring & Tutoring: Partnership in Learning, 24*(3), 201-212. https://doi.org/10.1080/13611267.2016.1222811

Estrada, M., Eroy-Reveles, A. and Matsui, J. (2018). The influence of affirming kindness and community on broadening participation in STEM career pathways. *Social Issues and Policy Review, 12*(1), 258-297. https://doi.org/10.1111/sipr.12046

Jay, T. M. (2003). Dopamine: a potential substrate for synaptic plasticity and memory mechanisms. *Progress in Neurobiology*, 69(6), 375-390. https://doi.org/10.1016/s0301-0082(03)00085-6

Kendricks, K., Nedunuri, K. V., & Arment, A. R. (2013). Minority student perceptions of the impact of mentoring to enhance academic performance in STEM disciplines. *Journal of STEM Education: Innovations and Research, 14*(2), 38-46

Lawlor, K.B., Hornyak, M.J. (2012). SMART GOALS: How the application of smart goals can contribute to achievement of student learning outcomes. *Developments in Business Simulation and Experiential Learning, 39*, 259-267.

Ong, M., Smith, J.M. and Ko, L.T. (2018). Counterspaces for women of color in STEM higher education: Marginal and central spaces for persistence and success. *Journal of Research in Science Teaching, 55*(2), 206-245.

Rockinson-Szapkiw, A., Wendt, J.L. & Stephen, J.S. The efficacy of a blended peer mentoring experience for racial and ethnic minority women in STEM pilot study: Academic, professional, and psychosocial outcomes for mentors and mentees. *Journal for STEM Education Research, 4*, 173–193 (2021). https://doi.org/10.1007/s41979-020-00048-6

Rodriguez, S. L., & Blaney, J. M. (2021). "We're the unicorns in STEM": Understanding how academic and social experiences influence sense of belonging for Latina undergraduate students. *Journal of Diversity in Higher Education, 14*(3), 441–455. https://doi.org/10.1037/dhe0000176

Merriam-Webster. (n.d.). Struggle. In *Merriam-Webster.com dictionary*. Retrieved June 1, 2023, from https://www.merriam-webster.com/dictionary/struggle

---

This program is supported in part by a grant to Virginia Tech from the Howard Hughes Medical Institute through the Inclusive Excellence Grant.

CHAPTER 12.

GUNPOWDER CODE CLUB

Bringing elementary school students' interests and passions into the classroom through coding

WENDY GIBSON AND KIMBERLY CORUM

ABSTRACT

Inspired by my fifth-grade students' passion and excitement when designing mathematics games using Scratch (a visual block-based coding language), I established the Gunpowder Code Club (GCC), an after-school coding club that meets weekly. Our goal for GCC is to give all students access to coding education. There are approximately forty upper elementary students from diverse backgrounds who regularly attend the weekly GCC meetings. During our meetings, students have time for free play to explore Scratch and other coding platforms (e.g., CodeMonkey.org, Code.org) to learn basic coding commands. They are also challenged to complete various coding tasks and projects. In addition to learning how to code, students are exposed to computational and algorithmic thinking, while further developing their mathematical thinking skills, problem-solving skills, and ability to work collaboratively. Students work together on projects, ask questions, and tutor each other while solving problems and debugging their code.

Studies show that girls begin to associate boys with science and math as early as second grade, and middle school is often when stereotypes and harmful associations cause many girls to avoid STEM subjects. The American Association of University Women (AAUW) reports girls and women are systematically tracked away from science and math throughout their education, limiting their access, preparation, and opportunities to pursue these fields as adults. In fact, women make up only 28% of the science, technology, engineering, and mathematics (STEM) workforce and men vastly outnumber women majoring in most STEM fields in college. The gender gap is particularly high in some of the fastest-growing and highest-paid jobs of the future, like computer science and engineering (Corbett & Hill, 2015). Racial discrimination is also prevalent in the STEM fields. As explained by Girls Who Code (n.d.), "Historical and institutional barriers—particularly racial bias and discrimination—play a role in the widening gender gap in computer science and who has access to opportunities in these fields." Underrepresentation in STEM fields can have long-term consequences for women and Black, Indigenous, and people of color (Pew Research Center, 2021).

INTRODUCTION

ESTABLISHING THE GUNPOWDER CODE CLUB

One of the most important goals of teaching computer science to young children is to foster the development of computational thinking skills that are applicable to many educational disciplines and areas of life (Barr & Stephenson, 2011; Chen et al., 2017; Cuny et al., 2010; Wing, 2006). All children, regardless of race, gender, or socioeconomic status, should have equitable opportunities to learn coding and develop computational thinking skills. The consequences of not providing equitable opportunities are dire. Researchers found that "…it is increasingly apparent that performance gaps by social class take root in the earliest years of children's lives and fail to narrow in the years that follow. That is, children who start behind stay behind—they are rarely able to make up the lost ground," (Garcia & Weiss, 2017).

An element of diversity, equity, and inclusion (DEI) excellence includes providing increasing access to coding education. During the Spring 2022 semester, I challenged my fifth-grade students to design a mathematics game appropriate for kindergarten students using Scratch (a block-based coding language) as a class assignment. We spent four days creating a mathematics game and then shared the game with a kindergarten class at our school. Fifth-grade students taught addition and subtraction operations to kindergarten students using the games they created. Student interest, motivation, and engagement with this project was so high that I went on to establish the Gunpowder Code Club (GCC) the following year. GCC is an after-school coding club that meets weekly. Our goal for GCC is to give all students access to coding education, thus leveling the playing field and providing opportunities for those traditionally marginalized in STEM education.

INCREASING STUDENTS' OPPORTUNITIES TO CODE

Evidence of our school's commitment to utilizing technology to support our students' education includes providing all our students a school-issued device, issuing students Scratch accounts, and providing initial instruction to coding with Scratch. Our school serves a diverse population of students. More than 40% of our students identify as Black, Hispanic/Latino, Asian, or multiracial. Nearly 28% of our students receive free/reduced price meals and nearly 20% of our students are classified as economically disadvantaged. GCC provides underrepresented minority students, low-income students, and those with little exposure or access to computer science and coding with new and exciting opportunities. Approximately 40 upper elementary students representing a range of diversity and ability levels regularly participate in the weekly after-school meetings. We have an equal distribution of girls and boys in our club and our school demographic draws from African-American, Caucasian, and South-East Asian populations. During these meetings, students explore Scratch basics, work together with peers to problem solve and experiment, and have fun playing and exploring. Students learn how to code in Scratch through free play and teacher-led instruction as they work to complete various tasks and projects, such as using Scratch to support game-based mathematics learning.

In addition to learning how to code in Scratch, GCC provides students with opportunities to develop computational and algorithmic thinking skills, while attending to mathematical practices such as engaging in rich problem solving and collaboration. Students work together on projects, ask questions, and tutor each other while solving problems and debugging their code. During each meeting, students are provided time to share their current projects or games on our large projection screen. Students come to the front of the classroom, plug their device into the teacher station, and discuss their Scratch project. Students have the chance to show their code, as well as what they have learned and how they have refined their projects. Students learn important skills from their peers related to problem solving, perseverance, and productive struggle.

The act of coding facilitates strong habits of mind for students in GCC. Giving students agency to demonstrate their learning through coding has highlighted students' depth of mathematical understanding while also generating excitement for learning and engaging in mathematics concepts and skills.

FOLLOWING STUDENTS' LEADS

There are many options for coding with a wide range of programming languages available to students. As the school year progressed, students were introduced to two additional code platforms to explore and expand their understanding and knowledge of coding. The next platform they explored was CodeMonkey.org. While CodeMonkey.org is a beginning programming language, this platform was more challenging and rigorous than most of our club members were ready to pursue, particularly for our fourth-grade members. The learning curve from Scratch to CodeMonkey.org was steep for some and required more teacher support. Two of our more advanced fifth-grade students did enjoy the challenge and spent time over winter break learning the basics of the programming language and successfully created simple games. However, given the choice of which coding platform to work with when students returned from their holiday break, all students returned to using Scratch.

I introduced the third platform, Code.org, halfway through the school year. Code.org has been a great fit for GCC and all students were very receptive to this platform. With just a few simple introductory activities and utilizing their prior knowledge from Scratch, students were able to complete various Code.org challenges at their own pace and create their own games. The most popular difference between Code.org and Scratch is the ability to send code projects to phones through a link or QR code. This allowed students to quickly and easily share their games with others. Introducing Code.org with its additional features and tools came at the right time for students, allowing them to build on their code knowledge from the beginning of the school year. Emily, one of our fourth-grade girls, reported that Code.org was more enjoyable because, "You can do more things." Sara, one of our fifth-grade girls, shared, "I like how creative we can get with it. It's not too complicated...I like how you can explore and learn and how you can code a lot more than on Scratch."

BUILDING COMMUNITY THROUGH CODING

GCC provides students with the opportunity to work in community to learn how to code. Students are often heard saying, "This doesn't work!" "Why is this not working!" and "I don't know how to fix this." As these frustrations are shared freely, other students eagerly respond with "Have you tried...," "I'll be right there," "Give me a minute and I'll take a look," or "I know how to fix that!" The collaboration of applied computational thinking and problem solving is on display as students share their works-in-progress, receive peer feedback, and add new features to projects as they learn more about the various capabilities in Scratch.

This sense of community can also be seen in the following vignette. During one meeting, Collin, a fourth-grade student, was excited to share his Scratch game with his fellow club members. In the midst of the demonstration, Collin realized his game character was not moving in the direction he intended and commented, "I see a problem I need to fix." During the demonstration, Donny (one of our fifth-grade club members) rose from his seat, approached the projector, and said, "I can help you. I think I see your problem." Donny proceeded to coach Collin through the debugging process by identifying an error in the directional code and explaining how it could be fixed. Collin was able to revise his code on the spot, tested his revised code, and saw that the game now worked as he intended. Collin was overjoyed that his code was working and returned to his presentation as if nothing had previously stopped him. This is just one example of the type of collaboration, computational thinking, and problem-solving that occurs each week in GCC.

STUDENTS' PERSPECTIVES

Based on responses to the Elementary Student Coding Attitudes Survey, students overwhelmingly report that "solving code problems seems fun," "coders are good at math," and "coders are good at language arts" (Mason & Rich, 2020). When asked to share if coding has helped them in school, students reported the following:

When asked to share if coding has helped them in school, students reported the following:

"Yes, I think it's because in Scratch you are a problem solver."

"Yes a little in math because there is a lot of thinking."

"Coding definitely helps me in school. My brain overtime improves how it functions which helps with math and other technology subjects."

When asked to describe how coding has helped them grow in other ways, students shared:

"It helped me grow my knowledge."

"Yes, because I am growing smarted in more than one way."

"It made me have new interests and I work at ideas and ideas make me grow."

Research on how children learn reveals that learning outside the classroom, if given the opportunity, often transfers to student engagement inside the classroom. As students pursue their interests, motivation increases, and often academic achievement is enhanced. Researchers have noted, "Students that engage in learning experiences outside of the classroom report having higher levels of motivation, recall the course material more vividly, and have improved academic performance in the class," (Claiborne et al., 2020). Fourth- and fifth-grade teachers in our school have reported students in GCC are showing greater interest in math class and are excited to share coding projects with other students. Students who participate in GCC have asked teachers' permission to work on code projects in class and have reported spending additional time after school and on weekends improving their code projects.

LESSONS LEARNED

Learning to code promotes problem solving and persistence. As students grow, progress, and mature through their mathematics education, they begin to discover the predictability and unpredictability of mathematics. Learning to code is a beautiful example of how to support students' conceptual understanding as their knowledge of coding continues to grow over the school year. Our goal with GCC is to give all students access to coding and opportunities to participate in rich mathematical thinking and problem solving, thus increasing student motivation and engagement in and out of the classroom. Learning to problem solve and persevere through productive struggle while coding develops strong habits of mind for students in GCC. Giving students agency to demonstrate their learning through coding has highlighted students' depth of mathematical understanding while also generating excitement for learning mathematics. In GCC, we integrate an inquiry-based curriculum with explicit instruction as needed resulting in transferable computer science and mathematics knowledge. Whether novice coders or experienced coders, the structure of GCC provides all students with the opportunity to create and share their knowledge of coding with a variety of student-driven projects. As a result, students have autonomy over their learning and are self-motivated to deepen their coding knowledge.

While there is a need for further research, our initial exploration revealed that the authentic learning experiences demonstrated by students in GCC provided insight into what motivates students. My students ignited around learning to code and their enthusiasm has been contagious. Students demonstrated perseverance and productive struggle accomplishments in ways I could only hope for in the classroom. Students were highly engaged as they were allowed multiple means of expression and choice. Mathematical interest and habits of mind increased during the various coding projects, as well as students' social and emotional well-being as they worked together and eagerly shared their learning with others.

There is a call to action to decrease barriers for marginalized students to participate in STEM fields. An advocate for STEM education, former First Lady Michelle Obama argues, "We need all hands on deck. And that means clearing hurdles for women and girls as they navigate careers in science, technology, engineering, and math" (STEM Like a Girl, n.d.). The Gunpowder Code Club is a proof-of-concept that providing students with opportunities to learn coding in a student-centered, inquiry-based environment results in transferrable skills that supports students' computational thinking, mathematical learning, and social-emotional well-being.

Acknowledgement

This project is based upon work supported by the National Science Foundation under grant No. 2243461. Any opinions, findings, and conclusions or recommendations expressed in this material are those of the author(s) and do not necessarily reflect the view of the National Science Foundation.

REFERENCES

Barr, V., & Stephenson, C. (2011). Bringing computational thinking to K-12: What is involved and what is the role of the computer science education community? *ACM Inroads, 2*(1), 48-54. https://doi.org/10.1145/1929887.1929905

Chen, G., Shen, J., Barth-Cohen, L., Jiang, S., Huang, X., & Eltoukhy, M. (2017). Assessing elementary students' computational thinking in everyday reasoning and robotics programming. *Computers & Education, 109*, 162-175. https://doi.org/10.1016/j.compedu.2017.03.001

Claiborne, L., Morrell, J., Bandy, J., Bruff, D., Smith, G., & Fedesco, H. (2020). *Teaching outside the classroom.* Vanderbilt University Center for Teaching. Retrieved June 7, 2023 from https://cft.vanderbilt.edu/guides-sub-pages/teaching-outside-the-classroom

Corbett, C., & Hill, C. (2015). *Solving the equation: The variables for women's success in engineering and computing.* American Association of University Women. https://www.aauw.org/app/uploads/2020/03/Solving-the-Equation-report-nsa.pdf

Cuny, J., Snyder, L., & Wing, J. M. (2010). Demystifying computational thinking for non-computer scientists. Unpublished manuscript in progress, referenced in http://www.cs.cmu.edu/~CompThink/resources/TheLinkWing.pdf

Garcia, W., & Weiss, E. (2017, September). *Education inequalities at the school starting gate: Gaps, trends, and strategies to address them.* Economic Policy Institute. https://files.epi.org/pdf/132500.pdf

Girls Who Code (n.d.). *Diversity, Equity, and Inclusion Statement.* https://girlswhocode.com/diversity-equity-and-inclusion

Pew Research Center (2021, April 1). *STEM jobs see uneven progress in increasing gender, racial and ethnic diversity.* https://www.pewresearch.org/science/wp-content/uploads/sites/16/2021/03/PS_2021.04.01_diversity-in-STEM_REPORT.pdf

Mason, S. L., & Rich, P. J. (2020). Developmental analysis of the Elementary Student Coding Attitudes Survey. *Computers & Education, 153.* https://doi.org/10.1016/j.compedu.2020.103898

STEM Like a Girl. (n.d.). https://stemlikeagirl.org/

Wing, J. M. (2006). Computational thinking. *Communications of the ACM, 49*(3), 33-35. https://doi.org/10.1145/1118178.1118215

This program is supported in part by a grant to Virginia Tech from the Howard Hughes Medical Institute through the Inclusive Excellence Grant.

CHAPTER 13.

USING DEPARTMENTAL BOOK CLUBS TO BRIDGE THE FACULTY-STAFF-STUDENT GAP

DEBORAH J. GOOD AND DEBBIE POLLIO

INTRODUCTION

Book clubs are one method for increasing discussion of topics in a safe place. In academia, they have been embedded in classes or used by faculty groups to discuss topics such as student learning. We describe a unique book club that is not course-based and includes faculty, staff, and students (graduate and undergraduate) from a department. All books had a diversity, equity, and inclusion (DEI) theme. We present an overview of the format of the book club, and data from a survey supporting its use to both increase a feeling of department inclusion among participants and increase understanding of DEI topics. Lessons learned for setting up a departmental book club are included.

BACKGROUND

According to one source, one of the earliest recorded book clubs was a women's religious reading club organized on a ship bound for the Massachusetts Bay Colony in 1634 (Overstreet, 2020). Fast-forward to 1996 and the start of Oprah's Book Club, which is thought by many to have caused a resurgence in book clubs among friends and neighbors. The "best-seller" type book club trend continues today, and now includes book clubs on Amazon.com and another (Reese's Book Club) led by Reese Witherspoon.

In academia, student reading is usually assigned, with either an in-class discussion or out-of-class writing assignment to follow. However, more course-based book clubs have been described in the literature, especially in psychology and sociology departments where using fictional accounts in popular books can help students engage with discipline-based topics (Khokhlova & Bhatia, 2023; Segrist & Meinz, 2018; Wyant & Bowen, 2018). In addition, faculty book clubs are becoming more popular as a way for faculty to engage outside of formal meetings or training in education topics (Rouech et al., 2022) or in diversity, equity, and inclusion-based books. Examples include a book club run by Radford University STEM (Science, Technology, Engineering, and Math) professors as part of their Howard Hughes Medical Institute (HHMI) grant work (S. Kennedy, personal communication), or those run by diversity, equity, and inclusion (DEI) offices, such as the Inclusive VT (Virginia Tech) Book Circles, of which this author (D.J.G.) is a participant.

It was following a meeting where the Radford University faculty book club was discussed and a subsequent discussion with a colleague who was doing a book club as part of her course that I (D.J.G.) decided to start up a departmental book club. I wanted the format to include faculty, staff, undergraduate and graduate students and even alumni, with a broad focus on DEI. A search of the literature did not reveal other clubs like it, in terms of the inclusion of all members of a department, and although some faculty-student book clubs do exist (i.e. (Khokhlova & Bhatia, 2023; Ney et al., 2023; Segrist & Meinz, 2018; Wyant & Bowen, 2018)), these exist as credit-bearing, not voluntary experiences, with only one exception found (Segrist & Meinz, 2018).

This chapter will describe this book club's format and provide data to support its use to increase a feeling of inclusion among members and increased understanding of DEI topics.

> After reading this chapter, individuals will:
>
> - Have the tools to start up their own departmental DEI book club.
> - Understand results and findings from a continuing 2-year DEI book club in a STEM department.

IMPLEMENTATION

Departmental list-servs and undergraduate student advising lists were used to announce the first meeting of the book club in Fall 2021. D.J.G. picked the first book, Educated, a Memoir by Tara Westover. This book has themes of religion/religious freedom, child abuse, and gender bias. The book was pre-selected by the organizer to ensure it was ordered in time for people to obtain it prior to the first meeting, and because the organizer had already read it and was prepared to lead the discussion. Ten books were ordered (approximately $150) with priority going to undergraduate and graduate students for the free book copies. Faculty and staff were encouraged to attend but were asked to obtain their own copy of the book. Meetings were set up using a poll for those who initially indicated they were interested in attending. Fourteen students (one graduate student), three faculty, and one staff member signed up for the first book club. Several individuals indicated that they had read the book and would not need a copy, and several people dropped out after the initial poll, so that ten books were sufficient for everyone who wanted a copy. This first book was long (352 pages, (Westover, 2018)), so meetings were set up every two weeks by web-conferencing (Zoom™) and in person to reduce the number of chapters for each meeting. As we were still in the COVID-19 pandemic, some members chose to join by Zoom, but most attended in person, wearing masks and socially-distancing. Of the seventeen originally signed up for the first book club, seven attended at least one of the meetings during Fall 2021. The list of book-club selections and number of members per semester is shown in Table 13.1.

Table 13.1: Book club selections and number of members by semester

Semester	Book Club Members	Book
Fall 2021	7	Educated, a Memoir by Tara Westover
Spring 2022	9	The Bluest Eye by Toni Morrison
Fall 2022	12	Being Heumann by Judith Heumann
		Disability Visibility by Alice Wong*
Spring 2023	12	Invisible Women by Caroline Criado-Perez**
		From Farmworker to Astronaut: My path to the stars by Jose Hernandez

[1]This book was added to the Fall 2022 list mid-semester and finished during the first meeting of the Spring 2023 semester because the members of the book club wanted more information on disability topics.

[2]This book was dropped at the first meeting of Spring 2023 because it was too data heavy. It was replaced by the Hernandez book mid-semester.

To increase participation, book club meetings lasting one to one and a half hours were usually held in the late afternoon/early evening, based on a poll of meeting times. The meetings included one meeting with food or at a restaurant to celebrate the end of the semester and pick the book for the following semester. While there was one book club that was held entirely on Zoom, it was decided that book-club meetings that were mixed (with some on Zoom and some in the room) were not effective, and these were discontinued the second year of the book club.

Figure 13.1: Advertising the Book Club. Survey respondents were asked how they found out about the book club. The results are shown in A. Most respondents found out about the book club through email, but several mentioned advertisements in class and the flyer. B. Two examples of flyers used to advertise the book club are shown. C. Bookmarks were created and handed out with each new book.

In 2023, a survey was sent out to all past members of the book club, as well as to all members of the Department of Human Nutrition, Foods, and Exercise at Virginia Tech. The survey was approved as "not research" by the Human Research Protection Program (protocol # 19-602). The purpose of the survey was to first, and second, to ask members about their experience in the book club, specifically asking about departmental inclusion, and the focus of the book club on topics of DEI.

RESULTS AND DISCUSSION

Thirty-two people responded to the survey, with 25% (8) of these being individuals who had joined the book club. Most individuals indicated that they had heard of the book club through email (75%), while 9.4% of respondents indicated that they had heard about the club through word of mouth. The remaining individuals had either heard about it through a flyer, or during a class when it was presented as an announcement. Email was the best way to reach all individuals in the department but including flyers with dates/times and the book for the semester seemed to reach some of the students (Figure 13.1A). All respondents indicated they were aware of the book club. Flyers were generated for each semester using online templates (Figure 13.1B). When the books were distributed, they came with paper bookmarks that had the dates listed and chapters due for the book-club meeting (Figure 13.1C). The bookmarks were made using a template in Microsoft PowerPoint, and printing on cardstock.

As previously shown (Table 13.1), membership in the book club reached twelve individuals. This is somewhat low for a department with nearly one thousand students but is a group size that is generally recommended for book-club meetings (Overstreet, 2020). Most of the individuals responding to the survey (75% of 32 individuals) indicated that they were not members of the book club at any time in the past four semesters. Answers to why people did not join the book club (open-ended question on survey), included a lack of time or scheduling conflict (fourteen), and not interested in the books (three) as reasons.

The eight individuals who indicated they were members of the book club identified the semesters during which they participated, with the breakdown being shown in Figure 13.2. An open-ended question asked the respondents to comment on why they joined the book club and why they stayed in the book club. Respondents (N=9) gave answers with two major themes: (1) to meet more people, and (2) because of an interest in reading in general and the topics specifically. These reasons for joining our book club are consistent with book-club membership in general (Overstreet, 2020). Survey respondents were also asked why they stayed in the book club, and responses included *"It is a good opportunity for networking within the major and I like having a distraction from school that still keeps me involved"* and *"I love the intimate atmosphere and the opportunities it has given me to get to know the faculty outside of the classroom. I've also connected with several of my peers and built relationships that go beyond the book club."*

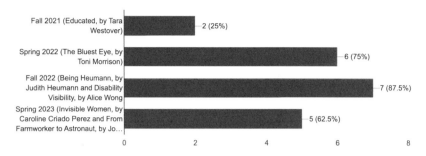

Figure 13.2: Survey respondents who are members of the book club. Survey respondents (8) who indicated they were members of the book club were asked which semester they were active in the club and were able to select more than one answer for this question

The book club was originally formed to provide a place to read and discuss books related to DEI topics, with a goal to increase awareness of DEI issues. A secondary goal was to break down barriers between faculty, staff, and students, providing a non-classroom environment to engage people with similar interests. As shown in Figure 13.3, most respondents overwhelmingly agreed that the club had met both goals, increasing their awareness of DEI topics, and increasing feelings of departmental inclusion. Interestingly, 9 individuals responded to this question, even though only 8 individuals had indicated they were members. The "n/a" responses may be from that 9th person. An open-ended set of comment questions followed, focusing on connection within the department. In terms of participants sense of connection, comments included *"I feel that my participation in the book club allowed me to feel more connected to the HNFE* [sic] (Human Nutrition, Foods & Exercise) *students and faculty who also participated in book club discussions because we have very open conversations and we learn a lot about each other and from each other"* and *"I have been able to connect with student and faculty I wouldn't have otherwise met"* and *"It shows students the other side of professors, connected by the joy of reading."*

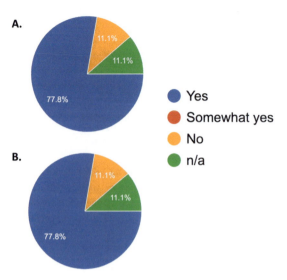

Figure 13.3: Book club themes. Survey respondents (9) who indicated they were members of the book club were asked (A.) if being part of the book club has increased your awareness of DEI issues and (B.) if being part of the book club has made you feel more like part of the department.

Table 13.2: Comments from the open-ended question, "which book impacted you most and why" (N = 7). The survey was conducted at the start of the fourth semester, so only the first three books were included in the responses.

Book	Comment
Educated by Tara Westover	• "Educated was the most impactful for me because I resonated with the author's experience without being familiar with her culture." • "I really respect how inclusion is an important and conscious decision for a happy community"
The Bluest Eyes by Toni Morrison	• "I read both "Educated" and "The Bluest Eye". I was incredibly impacted by both as I not experienced what the authors of both of those books experienced. These books changed my perspective in various but I believe what I received most was the conversations within the Book Club. To not only to my own perspective of the book but to also hear the perspective of others was so moving. I learned so much about not only life but also myself."
Being Heumann by Judith Heumann	• "I really enjoyed Judith Heumann's book because the disabled community is not usually the first to come to mind when thinking about disadvantaged groups. As such, my knowledge of the day-to-day for disabled people was limited but grew exponentially after reading Heumann's perspective" • "It impacted me by raising my awareness of the challenges (past, present, and future) that individuals with disabilities face every day as part of their lives. It also allowed me to better understand the process of lawmaking and policy work in the context of disability (and diversity, equity, and inclusion)." • "Being Heumann impacted me the most. I was aware of many of the issues discussed in the book, but reading and discussing the book made me understand disabilities and the history of legislation involving rights for those with disabilities on a deeper level." • "Being Heuman impacted me the most because disability issues aren't something I always think about as an abled person, and she has an incredible story."

We asked which book impacted them most, with participants' comments are shown in Table 13.2. The focus on DEI topics has clearly impacted group members' understanding and interest in diversity, equity, and inclusion. One person commented "*I firmly believe that these types of discussions can help us to become more introspective, purposeful, compassionate, and analytical, which in turn promotes DEI.*"

Overall, the outcomes of the book club were achieved, and the book club will continue in our department for the foreseeable future. Currently the club is funded using money from a grant from the Howard Hughes Inclusive Excellence fund, but our department is supportive of this project, and the costs are low (approximately $150 per semester). It is hoped that the transition from grant to departmental funding will be seamless. The current members regularly indicate that they are looking forward to continuing to use popular books to broach topics of gender, race, religion, sexuality, and inequity, among others in future semesters during book-club meetings.

Lessons Learned

Several lessons have been learned since starting the book club, and these are listed below, in hopes of helping others who want to form a departmental book club.

- It is important to have a lead for the club, and helpful to have another person to help. In this case, both authors have been consistent participants in the club from its inception. They help to organize and facilitate discussions, send out reminder emails, and generate flyers to announce the upcoming club topic. People interested in starting their own club might consider whether the lead for the club changes each semester, or continues with one person, and if so, how they are selected. The lead should be a faculty or staff member with access to departmental email list-servs.

- While we picked a different time each semester, it might be useful to have a consistent day/time for the book club each semester so that people can schedule it into their calendars. Student and faculty class schedules change each semester, but those who are interested in the book club (especially students) could plan around the timing of the club if they knew in advance. Faculty and staff may be more likely to come to the book club if it is not scheduled too late into the evening hours, conflicting with family obligations, dinner etc. We found that a 5 PM time, mid-week worked well for the members. We also initially had a meeting every two weeks (for the 352-page book that was picked in the first semester) but found that once-per-month meetings were sufficient for subsequent semesters.

- When choosing books, consider upcoming author visits to campus as this can also help to engage members. Judith Heumann's book was picked because the author was giving a zoom-talk the same semester.

- In choosing books, the survey respondents indicated that they liked the "democratic" approach, which makes sense in group dynamics. Respondents also liked receiving a copy of the book to read. However, in one case, a book was picked that no one had read (based on online reviews only), and it was a flop, with members choosing a different book mid-semester and not finishing that book. This increased costs for the semester and led to a discussion where members said they wanted at least one person who read the book before selection to avoid it again.

- The book club became a casual way to learn more about each other, and our interests. We also heard about other good books to read, since most people in the group were avid readers with several books on their side table. One of the comments from the survey stated "Personally, I did not have time to read an additional book. I would have loved to join a discussion based on short, current readings (like a journal club) because that would have been more convenient." In the future, we will make sure to emphasize that reading the books is never mandatory—in fact, many times the organizer (D.J.G.) had not yet finished all the chapters assigned. The topics discussed were based on the book, but many times diverged into personal stories or thoughts that everyone could participate in. It is important that there is no pressure to read the book and that everyone is welcome to attend the book club.

> **Reflective Questions**
>
> - If you were to start a book club in your department, would it be part of a course structure, or a separate non-graded club? What do you think the advantages/disadvantages are of these two distinct types of clubs?
> - What theme or topic would be the focus of your departmental book club? Are these themes structured for faculty-led discussions or inclusive of all levels of the department?
> - Who will fund the purchase of the books?

REFERENCES

Khokhlova, O., & Bhatia, A. (2023). Bringing books back: Enhancing the understanding of psychotherapy in psychology students through book club participation. *Teaching of Psychology, 50*(1), 32-40.

Ney, D. B., Ankam, N., Wilson, A., & Spandorfer, J. (2023). The implementation of a required book club for medical students and faculty. *Medical Education Online, 28*(1), 2173045. https://doi.org/10.1080/10872981.2023.2173045

Overstreet, M. (2020, October 16). I've got to talk to someone about this! A history of book clubs. *Book Riot.* https://bookriot.com/a-history-of-book-clubs/

Rouech, K. E., VanDeusen, B. A., Angera, J. J., Arnekrans, A. K., Majorana, J. C., & Brown, J. L. (2022). Read all about it: Faculty book clubs in action to support student engagement. *The Clearing House: A Journal of Educational Strategies, Issues and Ideas, 95*(6), 242-249.

Segrist, D. J., & Meinz, E. J. (2018). Looking for a good read? Running a psychology book club. *Psychology Learning & Teaching, 17*(2), 219-228.

Westover, T. (2018). *Educated: A memoir*. Random House.

Wyant, A., & Bowen, S. (2018). Incorporating online and in-person book clubs into sociology courses. *Teaching Sociology, 46*(3), 262-273.

To whom correspondence should be addressed: Department of Human Nutrition, Foods, and Exercise, Virginia Tech, Blacksburg, VA 24060; 540-231-0430; goodd@vt.edu

---

This program is supported in part by a grant to Virginia Tech from the Howard Hughes Medical Institute through the Inclusive Excellence Grant.

CHAPTER 14.

INSTITUTIONALLY ADVANCING INCLUSIVE EXCELLENCE

Leading from the middle in times of transition

JEANNE MEKOLICHICK; JAMIE K. LAU; AND SHARON BLACKWELL JONES

ABSTRACT

Like so many institutions of higher education, Radford University experienced significant leadership transitions over the past number of years. During the award period, leadership turnover in key institutional stakeholder positions occurred yearly, including—and importantly—at the provost and president levels. These repeated transitions, while challenging, created the opportunity for us to focus and hone our approach and strategies for advancing institutional change. Centering the whole student, and with attention to the contexts within which our students live, learn and thrive, foundational texts including Tinto's (1975) model of engagement and persistence, symbolic interactionism and identity theory (Mead 1934; Stryker; 2003), and intersectionality theory (Collins 1993; 2013) were at the core of our thinking. These theories suggest that proximate and intermediate campus environments that cultivate a sense of belonging and science identity can be critically important drivers of success for all students. More, we leveraged theories of organizational change to guide our strategies toward institutional shifts at multiple institutional levels as well as across units and divisions (Boyce 2003; Denofrio 2007; Golem 2018; Kotter 2018; Kuh et al. 2005, 2007; McNair, Albertine., Cooper, McDonald & Major 2016; McNair, Bensimon & Malcom-Piqueux 2020). We have been intentional, strategic and opportunistic in our efforts to alter the environments within which our students live and learn, attending to culture, structure and the political landscape. This chapter shares strategies employed, challenges encountered, successes enjoyed, and lessons learned in institutionally advancing our Inclusive Excellence (IE) initiative from the middle in times of transition.

INTRODUCTION

OUR APPROACH: CENTERING THE WHOLE STUDENT FOR ORGANIZATIONAL CHANGE

Like all the other IE projects, the goal of our program, REALising Inclusive Science Excellence (REALISE), was to effect positive cultural and structural change in our participating science, technology, engineering, and mathematics (STEM) departments and spark change to advance inclusive excellence across the college and the entire institution. With a focus on faculty professional development, curricular change, student programming, and institutional change, we aspired to build a community of empowered faculty and student learners and a university community that is student-ready, welcoming, and inclusive.

CHANGE FRAMEWORK

Our approach to achieving program goals was to center the whole student—attending to the contexts within which our students live, learn and thrive. We drew heavily on Tinto's (1975) model of engagement and persistence, symbolic interactionism and identity theory (Mead, 1934; Stryker, 2003), and intersectionality theory (Collins 1993, 2013). These theories suggest that proximate (classrooms and student groups) and intermediate (departments and colleges) campus environments that cultivate a sense of belonging and science identity can be critically important drivers of success for all students. We placed energy in cultivating our student groups, shifting student and faculty mindsets, and implementing inclusive pedagogical practices. As an overlay, we leveraged theories of organizational change to guide our strategies and approaches toward cultural and institutional shifts at multiple institutional levels, as well as across units and divisions (Boyce, 2003; Denofrio, 2007; Golem, 2018; Kotter, 2018; Kuh et al., 2005, 2007;

McNair, Albertine, et al., 2016; McNair, Bensimon, et al., 2020). We focused on capacity-building by mobilizing a critical mass of committed faculty, staff, administrators and students who were focused on advancing diversity, equity, and inclusion (DEI) at Radford University and leveraging a variety of available tools. At an organizational level, our approach to change was intentional, strategic and opportunistic.

With these goals and framework, we set out with a keen focus on implementation and seeding change. As a leadership team, our positions included Associate Provost, Dean/Interim Dean/Interim Associate Dean, several junior and senior faculty, and post-doctoral fellows. This diverse group allowed us access to different institutional levers, resources, and spheres of influence to keep advancing the work.

SELECT STRATEGIES FOR INSTITUTIONAL CHANGE

In this section, we share some strategies employed, offer examples of implementation and share how transitions influenced our direction or focus. We focus on areas that showcase how we leveraged the strategic plan, leveraged other institutional change, leveraged existing institutional structures and leveraged connections that built alliances.

LEVERAGING THE STRATEGIC PLAN

A foundational strategy initially articulated in the grant proposal was to invoke the strategic plan as a touchstone for advancing the work. The alignment between REALISE project goals and the University's new strategic plan has been a key factor in gaining the support of our institutional leadership and expanding REALISE beyond the founding three departments. Many of the goals and strategies put forth in the strategic plan aimed to remove barriers and promote student success. We identified three goals and specific strategies in our Strategic Plan: *Embracing the Tradition and Envisioning the Future 2018-2023* in the proposal and continued to forefront those strategies through the life of the grant up to the end of the Hemphill Presidency (under which the strategic plan was created). These included three goals focusing on creating opportunities for engaging students in experiential education, incentives to infuse undergraduate research in the curriculum, and faculty professional development. The work of the REALISE program advanced the work in each of these three areas. Once the presidency shifted, and as we began to move toward the expiration of the strategic plan, we focused less on these touchstones as motivating factors to advance the work because they became less powerful drivers for change.

LEVERAGING OTHER INSTITUTIONAL CHANGE

When the REALISE program was launching, we were also restructuring and revisioning our tutoring support and faculty professional development units. We were able to hire directors for both units and charge those units to advance the inclusive practices goals of the grant, as well as advance other DEI work as part of their work. This change was made possible because these two units reported to one of our grant co-authors. Further, in our audit of the tutoring program's practices and documents, we removed deficit language and intentionally hired student tutors that reflected our student population. Our faculty development and teaching and learning center infused best-practices in inclusive pedagogy into their approach and workshops. During this time both of these units were charged with developing a program or learning outcome focused on DEI (as did all of the other units that reporting to the Associate Provost). Over time, the tutoring unit was moved to another reporting line, although the focus on DEI remains. Faculty Development remains in the Associate Provost portfolio and continues alignment with project goals.

LEVERAGING EXISTING INSTITUTIONAL STRUCTURES

Thinking with the end in mind, we set out to infuse the REALISE work into existing institutional structures including in our tri-annual teaching and learning conference, Our Turn, hosted by Faculty Development/Center for Innovative Teaching and Learning, and in our annual Institutional Effectiveness Day (IE Day) hosted by our Institutional Effectiveness and Quality Improvement Office. With new Faculty Development leadership and a charge to advance DEI programming, our tri-annual teaching and learning conference began to include a DEI track. The number of DEI sessions as well as the number of participants in these sessions soared. In a second example, the provost at the time was investing in making Institutional Effectiveness a campus-wide signature event. During the two-year tenure of this Provost, DEI was infused in the breakout sessions and workshops of IE Day, and in the second year, we brought in Dr. Tia McNair, Vice President in the Office of Diversity, Equity, and Student Success and Executive Director for the Truth, Racial Healing, and Transformation (TRHT) Campus Centers at the American Association of Colleges and Universities (AAC&U), author of *Becoming a Student-Ready College,* and alumna of Radford University. During Dr. McNair's visit she delivered an interactive keynote and led a workshop. When this provost left, the annual IE Day event, having run for six years, ceased.

On the more opportunistic side of leveraging institutional structures, COVID brought significant numbers of faculty to our teaching and learning center. As the team infused inclusive pedagogies across their programming, the faculty coming for assistance to navigate teaching through COVID were exposed to inclusive pedagogies and the tools to implement them. In a second example, as we were nearing the end of the grant cycle, institutionally we were gearing up for our ten-year accreditation visit from the Southern Association of Colleges and Schools Commission on Colleges (SACSCOC) along with the associated Quality Enhancement Plan (QEP). A QEP is required for accreditation and is a comprehensive, five-year action plan that "reflects and affirms a commitment to enhance overall institutional quality and effectiveness by focusing on an issue that the institution considers important to improving student learning outcomes and/or student success." Several QEP proposals were reviewed as possibilities for the 2023-2028 cycle, and our team put forth a proposal using the data and concept of REALISE. Indeed, the timing of accreditation and request for QEP proposals afforded us a prime opportunity to expand our vision for inclusive excellence across the university. Our proposal was selected, SACSCOC approved, and the implementation of our vision is underway, indicating that inclusive excellence is important to and supported by both our regional accrediting board and Radford.In some cases, leveraging institutional structures were resistant to leadership change (e.g., QEP), others (e.g., IE Day) were not.

LEVERAGING CONNECTIONS AND BUILDING ALLIANCES

A key feature of any change initiative is capacity-building. We invested in relationships to expand the circle of committed faculty in the STEM College as well as building collaborations horizontally and vertically across Academic Affairs and institutional divisions. We were also awarded allied grants during this time which created the opportunity to amplify and expand our reach of the HHMI IE grant beyond the STEM college.

Each of the Core team members spent time promoting the program in the three departments and recruiting faculty to participate in the REALISE Faculty Learning Community. We focused on making in-roads with strategic institutional partners, including leadership in Human Resources and the Center for Diversity and Inclusion. We also promoted and hosted reading groups (sponsored by an HHMI Faculty Forums on Race Grant) inviting anyone who was interested to sign up, get a free book, join in discussions and develop action plans for change. These groups were led by our Diversity Educator and drew participants from across institutional divisions and campus sites including athletics, alumni relations, human resources, student affairs, and enrollment management.

Working across institutional divisions, an informal group of mid-level administrators and faculty with a shared goal of advancing diversity and inclusion, along with our Diversity Educator, started meeting to collaborate and collectively advance our work. We met monthly for over a year, traveled to the AAC&U DESS conference together, and had a paper accepted to present the following year, *You Are Here: Leveraging Resources, Building Alliances & Capacity for Change*, before being interrupted by the pandemic. We continued to meet and advance our DEI work, but the group dissolved with most of the participants leaving the institution. We also were fortunate to have representation from two REALISE leadership team members on our institutional Diversity Equity Action Committee (DEAC). During the award period, DEAC led a two-year effort to develop a strategic plan to advance DEI at Radford University. With REALISE members on this team, we were able to share our work and, at times, align efforts.

Unfortunately, the full yield from these investments was limited due to several institutional departures. While we had a core cross-divisional cohort of committed folks progressing through the reading groups together, when our Diversity Educator left, the cohort dissolved.

During the HHMI IE grant award period, we were awarded an HHMI Faculty Forums on Race Grant and a Jessie Ball duPont Fund grant that allowed us to augment the work of the HHMI IE grant beyond the STEM college to other departments, colleges, and institutional divisions.

Threaded through all four of these approaches, we incentivized people and activities, aligning with partner priorities and supporting their initiatives for our mutual gain. As personnel and associated priorities shifted, we redirected our efforts, energy and resources to other viable pathways to continue advancing the work.

CHALLENGES

Like all other institutions, leading change during the COVID years was marked by significant challenges—as well as extra stressors of advancing DEI initiatives amid global racial unrest. Radford University was no exception. As an added layer to the typical challenges navigated when implementing organizational change, the volume and pace of leadership transitions across the institution created an amplifying effect.

While not unusual to higher education institutions at this time, Radford University experienced successive leadership transitions that would persist across the award period. The number and pace of the leadership turnover in key institutional stakeholder positions was extensive and brisk. During the six-year period, we had three presidents, six provosts, three science deans, several Center for Diversity and Inclusion Directors and AVPs for Human Resources, and three IE program leads, experienced the departure of our sole DEI Educator, as well as lost two of the three grant authors and nine faculty (26% of those invested in REALISE) who participated in the professional development. These repeated transitions, while challenging, created the opportunity for us to focus and hone our approach and strategies for advancing institutional change.

Fatigue, particularly change fatigue, was felt across our institution and especially among folks who were working to advance DEI. As is typical, with every new leader comes a new vision and approach. Faculty, staff, administrators, and students advancing the work experienced frustration from investing in relationships and building trust that was continually erased and repeated as key stakeholder positions were vacated and new leaders arrived. An exemplar area was the AVP of Human Resources. The position changed multiple times over the past six years. The constant shifts in HR leadership, as well as personnel changes across the unit, significantly hindered our efforts to advance DEI initiatives in this critical partner area and ultimately led to investments of time and energy elsewhere in the organization where there was greater stability.

An adjacent thread was the shifting viability of strategic approaches that were no longer as beneficial or impactful after a transition. Connecting to the strategic plan and leveraging Institutional Effectiveness Day are two examples discussed above. In a more project-specific disruptive example, we had crafted training for the REALISE students, peer academic coaches/tutors, and students participating in our civic engagement program to include a social justice certificate training offered by our Center for Diversity and Inclusion. When the director left, we lacked the institutional expertise to continue the work, and had to find alternative experiences for students to replace this program. The new leader was not continuing the program.

The continual transitions forced us to focus and revisit our foundational documents regularly as we were continually tasked to introduce and garner support for the program from institutional leadership. These repeated transitions also forced us to think differently and creatively to continue the advancement of our efforts; we were simply unable to become complacent, being consistently pushed to critically review our strategies, tactics and approaches.

SUCCESSES

Amid the challenges, we are fortunate to have experienced a number of successes in implementing the REALISE program and seeding institutional change. Sometimes these successes were in spite of transitions, where we pushed through or pivoted. At other times, these successes were because of transitions, where we were forced to think deeply and critically about the work we were doing. No matter which path brought us here, we have advanced the inclusive excellence effort across the institution.

The faculty professional development we offer has shifted. And consequently, faculty have altered their courses, how they teach, and the mindsets they bring to student interactions. This shift is evidenced in data from our student surveys and focus groups, as well as in course syllabi and faculty artifacts.

As mentioned above, we were able to leverage the HHMI IE initiative to secure two additional DEI grants that have allowed us to expand the reach and depth of allied programming to advance institutional DEI work supporting faculty development and student success.

We have tangible evidence of change in institutional documents. Our Teaching and Research faculty hiring guide was revised from an inclusive lens and our faculty annual reporting system now specifically includes opportunities for faculty to share their DEI work. Further, a number of academic support areas (undergraduate research, Honors College, community and civic engagement, tutoring, global, and faculty development) include a DEI programmatic or learning outcome in their units.

We are seeing culture change with the creation of the Fresh Fruit Fridays (an initiative to engage faculty and students – see the chapter, Creating Impactful Moments of this book for details). This kind of event has been adopted across the institution in other colleges and areas, which now host similar events on separate days. Further, this event is included as programming for the QEP. These gatherings have turned into times when faculty and students gather and build community. Some faculty are even holding office hours during these times. Data from our faculty surveys and reflections indicate perspective shifts resulting from their participation in trainings and reading groups. With the creation and promotion of a plug-and-Play *Equity Gap Analysis Workbook,* by Sociologist Dr. Allison Wisecup, it is becoming more common for faculty to review disaggregated class-level data—and for faculty and departments to be discussing this information together.

In the end, with the selection of the QEP inspired by REALISE, we are thrilled to see key elements of the program live on at the institution beyond the three departments, the STEM college, and the life of the grant. This new program, REALising Inclusive Student Excellence (RISE), builds on the wisdom gained through implementing the REALISE program and has had positive structural and cultural impacts on faculty, student success and our institution overall.

LESSONS LEARNED

Like all projects, once we begin to implement them, new variables arise that direct change. In some ways, the shifts in our approaches during implementation are no different than implementing during times of leadership and stakeholder stability—and without a pandemic. In other ways, the continual flux forced us to work harder as we had to repeatedly introduce new leadership to the project and restart partner relationships. Of the many lessons learned from seeding institutional DEI change from the middle, we focus on three.

One powerful lesson we learned was the immensely positive impact of having an experienced, well connected, expert Diversity Educator, Dr. Sharon Blackwell Jones, on our campus. In addition to the many DEI trainings and book groups she led, she spent numerous hours visiting with faculty helping them implement what they learned and process and navigate the challenges that ensued, as well as supporting our students of color individually and in groups as they navigated our primarily white institution with an emergent DEI culture. The loss of her powerful presence on campus and our inability to secure a position to rehire behind her, was felt across the institution. While gains can be made through consultants and virtual workshops, having a person in place who can build relationships, hop on a call, or show up to help in a class has significantly greater impact. We used this lesson to advocate for hiring a Diversity Educator at the beginning of our QEP to set us up for a successful implementation.

In light of the continual leadership transitions, we learned how to help one another see the positive impacts that we have made and continue to make in our areas. We were able to see how, collectively, those of us in the middle were affecting change by honing our focus on our sphere of influence and pressing forward. Helping to center the power we have within our spheres of influence and the coalition built, aided in reenergizing the group to keep going. Inclusive excellence is hard, emotionally laborious work and supporting one another is critically important to continued engagement in the work.

We learned that there are always paths forward to advance inclusive excellence work. Adopting a positive, proactive, and nimble mindset, ready to shift focus from structure to culture, from process to people, from one area of the organization to another, learning when to pause and when to push forward, helped keep the movement—and the people—going.

Acknowledgement

Sincere thanks to the Howard Hughes Medical Institute's Inclusive Excellence in STEM Education Grant (#52008708) for funding our REALISE program, all the faculty and students who participated in the REALISE program, and all the administrators, faculty and staff who invested in seeding institutional change.

REFERENCES

Boyce, M.E. (2003). Organizational learning is essential to achieving and sustaining change in higher education. *Innovative Higher Education, 28*(2), 119-136.

Collins, P. H. (1993). Toward a new vision: Race, class, and gender as categories of analysis and connection. *Race, Sex, and Class, 1*(1), 25-45.

Collins, P. H. (2013). *On intellectual activism*. Temple University Press.

Denofrio, L. A., Russell, B., Lopatto, D., & Lu, Y. (2007). Mentoring: Linking student interests to science curricula. *Science, 318*(5858), 1872-1873. https://doi.org/10.1126/science.1150788

Golom, F. D. (2018). Alternate conversations for creating whole-system change around diversity and inclusion. *Diversity & Democracy, 21*(1), 16-19.

Kotter, J., (2018). *The 8-step process for leading change*. Kotter. https://www.kotterinc.com/8-steps-process-for-leading-change

Kuh, G. D., Kinzie, J., Schuh J. H., Whitt, E. J. & Associates. (2005). *Student success in college: Creating conditions that matter*. Jossey Bass.

Kuh, G. D. et al. (Eds.) (2007). Piecing together the student success puzzle: Research, propositions, and recommendations. *ASHE Higher Education Report, 32*, 1-182.

McNair T. B., Albertine S., Cooper M. A., McDonald N., & Major T. Jr. (2016). *Becoming a student-ready college: A new culture of leadership for student success*. Jossey Bass.

McNair, T. B., Bensimon, E. M., & Malcom-Piqueux, L. E. (2020). *From equity talk to equity walk: Expanding practitioner knowledge for racial justice in higher education*. Jossey-Bass.

Mead, G.H. (1934). *Mind, self, and society from the standpoint of a social behaviorist*. University of Chicago Press.

Stryker, S. [1980] (2003). *Symbolic interactionism: A social structural version*. Blackburn.

Tinto, V. (1975) Dropout from higher education: A theoretical synthesis of recent research. *Review of Educational Research, 45*, 89-125.

This program is supported in part by a grant to Virginia Tech from the Howard Hughes Medical Institute through the Inclusive Excellence Grant.

CHAPTER 15.

FROM THE SOUL

Learning and Leading Together toward Inclusive Excellence
CYNTHIA A. DEBOY; LAURA GOUGH; SARAH A. KENNEDY; AMANDA C.
RAIMER; AND JILL C. SIBLE

ABSTRACT

From the first gathering of the Inclusive Excellence Cohort 1 (IE1) program directors (PDs), the Howard Hughes Medical Institute (HHMI) challenged us as project leaders to think and act differently in order to make real progress toward equity and inclusion at our institutions. In the midst of grappling with the details of our individual projects, we were quick to embrace the mantra "fix the institution, not the student" and to commit to leading change in curriculum, pedagogy, and even culture. At the time, we did not anticipate the profound impact that a commitment to inclusive excellence would have on us personally and professionally. Here, five leaders from four of the IE1 institutions reflect on how we adopted a learning mindset and built a community and friendships that enabled us to lead from the soul even during the darkest hours of the pandemic and critical period of racial reckoning.

INTRODUCTION

Leading change – it's hardly a novel topic. There are hundreds of books, courses, and countless theories to support the work. Even through the narrowed lens of leading change in academia, resources abound. When we applied for grants to be in the first cohort of institutions to "engage in the continuing process of increasing their institution's capacity for inclusion of all students," we understood that we were being called to adopt a different approach and mindset, to stop trying to fix perceived deficits in our historically excluded students, and instead, to lead deep and sustained change in the ways in which our institutions engaged with our students. Not just meeting them where they were but understanding and embracing where they had come from. Not creating new programs targeting those students our institutions labeled "at risk" and "underprepared," but dismantling and rebuilding our classes and curricula so that all students encountered a pathway to success. We understood the charge and were motivated to begin. But did any of us actually know how to be the leaders we needed to be to move our campuses toward inclusive excellence? Not at all. We honestly felt quite overwhelmed and humbled. Had we the foresight that this work would take place during a pandemic and a critical period of racial reckoning following the murder of George Floyd in 2020, we undoubtedly would have been even less confident in our capabilities as leaders. Fortunately, the Howard Hughes Medical Institute (HHMI) supported us in working together as a community and adopting a learning mindset; here we delve into how these two aspects of the Inclusive Excellence (IE) program allowed us to effectively improve the diversity, equity, inclusion, and justice (DEIJ) culture at our institutions.

But first, a little about who we are. As we have summarized previously (Wojdak et al., 2020), our institutions differ greatly in size, student demographics, and Carnegie classification, and at first, we questioned to what extent we would find common ground. As individuals, we hold quite different positions at our institutions and have varied levels of experience ranging from postdoctoral fellows to administrators close to earning free parking at their institutions. We share several aspects of our identity (white, female, natural scientist) that likely provided some common points of reference. Below (in the boxes), we share a little more about the selves we brought to this leadership circle.

FROM COMMITTEE TO COMMUNITY

The most important tool that HHMI equipped us with as leaders was each other. Beginning with the first program director (PD) meeting in 2017, institutions were organized into Peer Implementation Clusters (PICs) based on relative geographical proximity. HHMI encouraged us to work together and provided additional funding earmarked strictly for PIC activities. We were fortunate that the distances among our campuses were drivable, affording the opportunity for in-person annual meetings before the pandemic (more on that later) and more frequent iterations in a pair-wise manner (Trinity Washington-Towson and Radford-Virginia Tech). The experiences of IE grantee institutions with their PICs varied greatly across the country. Some found value in the partnerships; some did not. For us, the PIC and especially the relationships that developed among the program leaders were essential in fostering the areas where we moved the needle toward inclusive excellence on our campuses, and in building the resilience to move beyond challenges and failures, both personal and professional. Here, we tell the story of our leadership circle in the hopes that we might help others create this asset for their own work.

At the first IE meeting in August 2017, each principle investigator (PI) plus one additional representative from each institution met to plan our first meeting the following spring. We discussed how we would need regular meetings to plan the logistics of the meeting, but at the time, did not envision anything more than a working group, and only held a couple of meetings that year that were haphazardly scheduled. In the second year of our grants, our PIC Assigned Liaison (PAL) at HHMI requested a regularly scheduled monthly Zoom meeting with the four PIs.

Those monthly meetings started without any set agenda, but naturally evolved into an opportunity for each PI to update the group on project progress and challenges. We also used that time to ask questions of our HHMI PAL. Because each of us were in different faculty and administrator roles at our distinctive institutions, we often shared the particular issues we were facing to receive feedback from each other. These issues frequently extended beyond the bounds of our respective IE projects and reminded us that despite our different project approaches and institutional characteristics, we were all facing some of the exact same challenges: e.g., faculty buy-in, administrative support, incorporating student voice.

While we never called ourselves a community of practice (CoP; Wenger, 1999), we now recognize ourselves as such. Our commitment to our meetings and each other was crucial to building trust and vulnerability. Perhaps even more fitting than community of practice, is the term, "circle of empowerment" coined by Rita Irwin (1995) in a case study analyzing a leadership group of women arts educators in Canada. Like that group, we share power, mentor each other, and bring our whole selves to the circle. Our PIC leadership group also shares characteristics of the "Learning Circles" of school administrators in Southern Australia. Like our PICs, these circles were structured by the host organization to bring together school leaders around a common change initiative (Peters & Le Cornu, 2005). One key difference is that the Australian learning circles were more structured and facilitated compared to our PIC, where we were largely left to our own devices to self-lead.

One of our greatest strengths as a group lies in the diversity of our institutions, our projects, and our roles within those institutions. Each of our universities is vastly different, making it easy for us to not see each other as competitors but instead as supporters and co-conspirators. Meanwhile, our various leadership and DEIJ experiences were extremely valuable in providing support and helping one another brainstorm and problem-solve; finding our common ground led to creative fusions of ideas and innovations. This diversity made it even more powerful when we found threads across all of our programs that were successful, such as bringing faculty together in informal reading groups learning about DEIJ and higher education.

Reflecting on the past six years, we realize that we somewhat organically adopted the community norms for our CoP that our DEIJ professional colleagues use in our inclusive excellence workshops at our individual institutions and PIC meetings. These include listening deeply for understanding, participating with kindness and respect, and remembering that what is learned in these sessions can be shared, but the specifics of the conversation are private. This is one of several examples of how the activities we engaged in to move towards IE in one setting influenced how we participated in similar activities in different settings.

A COMMUNITY OF LEADERS... AND LEARNERS

The second tool HHMI instilled in us was evoking a learning mindset: most of our annual PD meetings engaged us in professional development to improve our competencies related to DEIJ. The HHMI program directors emphasized that they were learning right along with us. We were told to consider our projects as experiments and were given latitude to pivot when something failed or an unanticipated opportunity arose. Our annual reports, called PIER (Progress toward Inclusive Excellence through Reflection), were like no grant reports we had ever written. They were purely reflective with no figures or tables allowed. We came to appreciate that HHMI valued what we had learned even more than what we had done.

Interactions with our DEIJ professional colleagues reinforced the idea that we are all continually learning in this arena. Our CoP became a space in which we could all learn from each other, continuing our role as learners while leading. Perhaps none of us realized at first how expansive this project was intended to be until we were tasked with transforming our institution for inclusive excellence. We were only beginning our journey to understand what inclusive excellence meant, let alone knowing how to transform an institution! This endeavor required navigating the workings of our institution, working with administration, faculty, staff, and students, and applying a systems-based approach. In the learning circles described by Peters and Le Cornu (2005), the academic leaders were similarly situated as learners. While our charge was less explicit, we found ourselves adopting this role as well.

Month after month and year after year, we met (usually online) and learned from the approaches we each took at our own institutions. By knowing the elements of each other's projects, we could ask for advice or resources, such as effective workshop facilitators or assessment tools.

We were also able to hold each other informally accountable as we shared accomplishments as well as goals yet to be met and provided space for reflection and discussion. Because of the safe environment we created by being open about challenges, the CoP also became a sounding board for new ideas and approaches. This allowed us to have a broader perspective for the approaches we were implementing to increase inclusion. We gained insight into the successes and challenges of various approaches at different types of institutions, which guided our own approaches toward solutions; with a mix of roles within our CoP we were further able to explore different ways to initiate change for faculty and administrators. Importantly, we recognized the similarities of challenges and how these often had root causes coming from a broader context within the culture of academia and society. By seeking to bring institutional change to not only one institution, but collectively at our four institutions, we aimed to make a small but meaningful shift in ideas, practices and culture and awareness toward greater inclusion. The value that we each placed on the importance of this work added to our cohesion and the need for support among the group. To lead inclusive excellence projects requires us to push and dismantle institutional norms, which requires confidence, perseverance, and courage. The support from the group and knowledge that this type of work was being conducted at other institutions was inspiring and encouraging. A monthly opportunity to celebrate progress and successes was also critical in keeping us motivated.

AND JUST WHEN WE THOUGHT WE KNEW WHAT WE WERE DOING...

...2020 happened, and along with the rest of the world, our IE projects, our institutions, and our lives were disrupted. Fortunately, we had each other. When we went into pandemic lockdown, we continued our monthly meetings which allowed us to share how our institutions were handling the situation, share resources, and acknowledge how difficult everything had become for our institutions and particularly for our students.

Together, we recognized the ways in which we were able to bring an inclusive lens to bear on how our departments and institutions were responding to moving all instruction online, and this monthly dialogue helped us consider other approaches and student and faculty needs. It was also at this time that our meetings began including more personal elements, partly because of the incredibly stressful situation of being in lockdown.

In 2020, not only was our CoP a source of professional support to maintain our IE focus in the face of an unprecedented pandemic, but our CoP also created support after George Floyd was murdered. By that time we had been meeting monthly for two years, and we had established an atmosphere of trust and vulnerability that made it easier to be open about the difficulties we were facing personally and professionally. We discussed the ways in which our projects were or were not addressing structural racism at our institutions and how we might become better equipped to do so.

It was also around this time that the PDs realized we needed to meet without a representative from HHMI. Although we appreciated the desire of HHMI personnel to be learning partners with us, because they provided the funding for our projects and we were reporting back to them annually, the power dynamic prevented the PDs from being entirely open. We continue to hold monthly PD meetings now, during the sixth year of our projects.

In our IE professional development workshops, we have spent a great deal of time discussing the "whole student" and how we can't ignore a student's life outside of our classroom or laboratory. In our CoP, we began to embrace the analogous situation of discussing the "whole faculty member" and recognizing that the pandemic, racial reckoning, and our personal lives were affecting our ability to accomplish our project goals and affecting the faculty and staff who work on IE projects with us. As to be expected in any group such as ours, life events also occurred during these six years to members of our CoP: a new baby, a retirement, death of a parent, spouse illness, along with the challenges and joys of parenting. We shared these experiences and gained support that was critical for our own emotional and mental health so that we could continue with our IE work. While our meetings remain agenda-less, we organically create space for sharing our personal and professional joys and challenges as well as the business of our shared work.

CONTINUING AND EXPANDING THE COMMUNITY

Beyond our monthly meetings, we seek each other out at conferences and gather socially. Our larger PIC community convenes for an annual conference-style meeting, with each institution taking a turn as host. We also get a lot done together. We have published together and co-presented at conferences (e.g. Transforming STEM Education hosted by the American Association of Colleges and Universities). As a group, we mobilized collaborative projects among our universities. Through our yearly PIC meeting, we provide opportunities for our faculty to mingle with each other and build cross-institutional relationships. Two collaborative projects stemming from this work include dissemination of IE work at American Chemical Society meetings and the creation of a VT/Radford STEM inclusive pedagogy book group. In our final years of the grant, our PIC is working together to disseminate our work (in this collaborative book) and also to create avenues for our students to present their undergraduate research.

WHAT'S THE SECRET TO A GREAT COP?

Unlike a traditional research grant, the HHMI IE grants were intended to transform people and institutions. This work demanded that individuals reflect on their teaching, disaggregate their student success data, and be honest about who their programs are best serving. While the most tangible artifacts of this work at institutions might be course-based undergraduate research experiences (CUREs), project-based learning, or inclusive curricula, the hardest work was encouraging faculty to question their assumptions, adopt a growth mindset toward students, and take ownership of their role in removing barriers to student success.

Our leadership CoP was critical for this most difficult part of the work. Why did this group become so essential? What was the magic? We believe the key elements to its success include:

- A shared purpose and passion for the work. Each of us viewed the work of Inclusive Excellence as essential, not just at our individual institutions but across higher education.

- Time together. We committed to our monthly meetings, made additional time to support one another as needed, and sought opportunities to gather.

- Our diversity. The differences in our roles and institutions gave us a range of perspectives and positioned us well to collaborate, rather than compete.

- A safe space. We built a leadership circle where it was OK to share our failures, frustrations, and fears. We built one another up during the most difficult times.

- Joy! We remembered to celebrate the wins, personal and professional. We ate together (sometimes virtually) and laughed often.

This group was exactly what was needed to help move all of our institutions forward towards inclusive excellence.

Jill

While I joined this project as a seasoned administrator with experience leading other STEM education grants, I can't say that all those years translated into certainty about how to create wholly equitable experiences for our students. I serve as the Associate Vice Provost for Undergraduate Education at Virginia Tech. I hope that my roots as a first-generation college student and a faculty member (Professor in Biological Sciences) have kept me grounded in ways that always center students in our work and hold faculty with the utmost regard for the many hats they wear and the ways in which many put their hearts and souls into

Jill

their teaching and mentorship of our students. One of the greatest "aha!" moments for me over the course of this project has been the realization that **if we are to support faculty in adopting a growth mindset toward our students, then we must do the same for them.**

Our PIC leadership community of practice has provided an unanticipated space for me to extend that growth mindset to myself as an academic leader. My peers have provided the support and encouragement to bolster me to be a little bolder in my DEIJ work and more resilient in the face of opposition and setbacks. I seek to become the servant leader that Dan Cable describes as possessing "confident vulnerability" (Cable, 2023), although I admit that I am still working on the confident part of that equation. My PIC colleagues have helped with that. And beyond the professional benefits of our "circle of empowerment," I have made cherished friendships, for which I am deeply grateful.

Sarah

Unlike my co-authors from VT, Towson, and Trinity Washington, I was not the PD from the outset of Radford's Inclusive Excellent grant and was not at Radford when the grant was written. I was invited onto Radford's IE project's (REALISE) leadership team in the grant's first year (2018) while I was an Assistant Professor and part of the first Faculty Learning Community cohort. I attended AAC&U's Project Kaleidoscope 2019 STEM Leadership Institute to solidify my leadership skills and support my growth in the inclusive excellence area. Little did I realize, the groundwork of this PIC CoP was being laid as I was beginning my IE leadership journey and it would become a critical support structure for me. In Spring 2021, during my first year as an Associate Professor, our REALISE Program Manager transitioned to another university and asked me to step into grant management. A few months later, our Program Director was appointed as Interim Provost and asked me to take the helm of the REALISE program. In my new role as PD, I was overwhelmed as I didn't have experience leading a large STEM grant. However, having familiarity with Jill, Laura, and Cynthia from the annual PIC meetings, I was able to join in their monthly PD virtual meetings with a level of comfort, knowing them as thoughtful and caring women STEM IE leaders. These monthly meetings helped me combat imposter syndrome and feel confident that I was the right person to lead REALISE at Radford. In the second year of leading REALISE, Amanda Raimer took the helm as Program Manager and while I was on maternity leave in Fall 2022, Amanda joined the supportive CoP network of the PIC program directors. Jill, Laura, and Cynthia provided wonderful examples of how to lead effective teams and sparked ideas about how to engage our leadership team and faculty. Our CoP was a safe space to share about challenges and get critical feedback that would enhance our programs. I was welcomed back from maternity leave and my daughter made a few appearances at our meetings too!

Inclusion work in higher education requires us to examine inequities, understand and embrace diversity, and advocate for removal of barriers to student success. Specifically, following the murder of George Floyd, I delved deeper into inclusive excellence and anti-racism literature to examine systemic practices in higher education that exclude Black students. The CoP became a space to have difficult conversations and to provide support to each other as campus leaders who are advocates for our vulnerable students. Having shared identities as cis-gendered, white women STEM leaders, we recognized our critical role as allies, advocates, and change agents. This CoP equipped me with knowledge, confidence, and a safety net; it allowed me to be brave in DEIJ space. At Radford, under my leadership (even through battling imposter syndrome), the REALISE Program is becoming institutionalized as our university-wide Quality Enhancement Plan RISE: Realizing Inclusive Student Excellence.

Amanda

As Sarah mentioned, I'm the "newbie" and came into this wonderful group in a roundabout way. I joined Radford University and the REALISE Program in Summer 2020 as a Postdoctoral Teaching Fellow in biology. My main role was to provide reassigned time for faculty going through our learning communities and also help with various aspects of the REALISE grant. While I had some training in teaching, I had little experience with DEIJ work and saw myself as solely a learner in this space. My time with REALISE helped me build more confidence in incorporating

Amanda

DEIJ into my teaching, classroom, and other interactions with students, faculty and staff. I could see myself as a mentor for students but did not feel prepared to be a leader. Being one of few postdocs on campus and newer to this journey left me unsure of my capabilities.

As fate would have it, my role in the REALISE program would change drastically over a short period of time and challenge my perceptions of myself as a leader. After two years at Radford, I became the program manager for the grant, and then within a few months became the interim director while Sarah was on maternity leave. I was already hesitant about assuming the role of program manager since I felt I had little to no background in that skill set, but then to add on the PD position had me feeling completely out of my depth. So when Sarah invited me to my first directors meeting, I was very nervous. Here was this group of women who in my eyes were (and still are) absolute powerhouses in this work, while here I am just beginning to really find my footing. Talk about some serious imposter syndrome! However, within the span of one meeting, this group turned from something I was dreading to something that I look forward to every month. I quickly saw that these women were not only peers, but friends and true allies in this work that were happy to welcome me to their team. They clearly took the time to learn each other's projects and offered valuable ideas and feedback from their various backgrounds and experiences. They also celebrated each other's successes and supported each other through challenging times, both professional and personal. **Their vulnerability with each other helped me see that we were all learning, making mistakes, and growing at some level, which in turn has helped me feel more comfortable in my role as a leader.** Even though Sarah is back in the director's seat, I still attend the weekly meetings; these ladies are an invaluable group of mentors and our conversations fill my cup every time.

Cynthia

At Trinity, our project developed from the vision we created as a team of five. My role as a leader was to incorporate ideas from each person to create a cohesive, achievable project and foster a unified team. Each member from our diverse group brought strengths and passions that contributed to the project impact. We developed an identity as a STEM team working towards inclusive excellence. We participated in many PD workshops together and applied what we learned to revise our science program. We explored topics including implicit bias, positionality, combatting microaggressions, cultural competency, and learned about ourselves and each other. We were committed to be become better for our students, ourselves and each other. The process of learning together and incorporating ideas from everyone within our team meant listening and recognizing each member's valuable input and experiences. Working as a team meant working through challenges. We used these experiences through open and honest dialogue to become more inclusive in our communication, a stronger team, and better for our students. Working as an effective team took cooperation, concession, collaboration and sometimes apologies, self-reflection and moving on, all while learning from the experience. Most importantly however is that we value each other and the contributions we each make.

Cynthia

As our team grew as additional faculty members joined, I aimed to align each person's interests with a meaningful project goal to maintain cohesion. Integral to our team identity and capacity to make change remained acknowledging the value and contributions each person brought to the process. The practice of informal celebration contributed to pride for being part of this team. For example, faculty might share with each other about an impactful class session and the team would affirm and share in the success.

Inherent in our team's successes making change is the value we have placed on our team's diversity and recognition that individually we each provide value, while together we create a synergy to impact changes towards inclusive excellence and positively impact students.

In addition to the Trinity team, connections made with leaders from our PIC institutions became a supportive community. Our PIC community became a space in which I could step away from a narrower lens of norms at my own institution to develop perspective for the work, successes and challenges I was encountering from a broader vantage point. Each member of this PIC leadership community provided understanding, encouragement, problem solving strategies, and compassion with nonjudgmental support. From these meetings, for example, I learned about professional development opportunities and strategies for pivoting to online and hybrid teaching for science labs during the Covid pandemic. Our meetings therefore provided practical advice, inspiration, perspective and recharging for working on our project at Trinity.

Leadership on this project has been in large part about contributing to developing supportive communities both at Trinity and within the PIC. Leading has been about learning and working with teams of people with similar goals who when working together have vast experiences and knowledge to effectively create transformative change towards inclusive excellence.

Laura

I joined the Department of Biological Sciences at Towson University (TU) as Chair and Professor in August 2015. Shortly after I arrived, a colleague told me about the HHMI IE proposal he and others had begun to develop and asked if I would lead the project. From my previous institution I was already familiar with the pedagogical approach for the proposal and quickly agreed. One reason I accepted the position at TU was to be able to make change for a large group of diverse students, and this proposal fit right into that goal given the demographics of TU's student body. Although I had a track record of federal funding to support my plant ecology research, I had never run a STEM education focused project before. As we began our professional development sponsored by HHMI and our own faculty professional development for our program, I found myself continuously thinking about how what I was learning could be applied at the level of my department and potentially be scaled up to our college. As a department chair, I was able to implement new approaches to curriculum and advising for the entire department, not just the faculty participating in our HHMI IE program. I brought data and summarized workshops at our college-level leadership meetings to share with the other STEM chairs and our Dean. I also had contacts throughout the university whom I could readily engage in discussions such as the director of the Career Center, our Vice President for Inclusion and Institutional Equity, and our center for faculty teaching and learning because as a chair, I was already interacting with them in other ways. Although leading a project like this while chair was challenging, being in this position allowed me to **leverage my administrative role as department chair to further IE as approaches I learned through our IE workshops and activities applied to everything we do as a department**. And through it all, the CoP supported me. Leading a large science department through the pandemic lockdown (and at the same time a move into a new science building) was brutal, yet every month I was able to talk with the CoP members to hear how their institutions were handling the situation and again, I could leverage my position by bringing their ideas to our college leadership. And as I faced personal challenges over the past six years, the CoP members listened and offered support. Every DEIJ professional I have met describes how difficult this work can be, how slow it is, and how the rewards and successes feel incremental and far between. Having the regular meetings of the CoP and the trust and support of the group has helped me recognize the successes for myself and also mark them with my colleagues at TU so that we are reminded that we are making progress in the right direction.

REFERENCES

Cable, D. (2023, January 23). Confident vulnerability: Three ways for leaders to inspire others. *Forbes*. https://www.forbes.com/sites/lbsbusinessstrategyreview/2023/01/23/confident-vulnerability-three-ways-for-leaders-to-inspire-others

Irwin, R. L. (1995). *A circle of empowerment: Women, education, and leadership*. SUNY Press.

Peters, J. & Le Cornu, R. (2005). Beyond communities of practice: Learning circles for transformational school leadership. In P. Carden & T. Stehlik (Eds.), *Beyond communities of practice: Theory as experience* (pp. 107-132). Post Pressed.

Wenger, E. (1999). *Communities of practice: Learning, meaning, and identity*. Cambridge University Press.

Wojdak, J., Phelps-Durr, T., Gough, L., Atuobi, T., DeBoy, C., Moss, P., Sible, J., & Mouchrek, N. (2020). Learning together: Four institutions' collective approach to building sustained inclusive excellence in STEM. In K. White, A. Beach, N. Finkelstein, C. Henderson, S. Simkins, L. Slakey, M. Stains, G. Weaver, & L. Whitehead (Eds.), *Transforming institutions: Accelerating change in higher education*. Pressbooks. http://openbooks.library.umass.edu/ascnti2020/chapter/wojdak-etal

---

This program is supported in part by a grant to Virginia Tech from the Howard Hughes Medical Institute through the Inclusive Excellence Grant.